Mediation of Construction Disputes

This book is dedicated to David Miles, friend, mentor, encourager and advisor, to whom nothing is too much trouble and who was there at the beginning, is now and ever shall be the epitome of a generous man;

And to Lawrence Kershen QC who is the nearest thing that I could ever get to having a brother and who, beyond his seemingly endless compassion, has consistently reminded me that life can be FUN.

Mediation of Construction Disputes

David Richbell

Blackwell
Publishing

Blackwell Publishing editorial offices:
Blackwell Publishing Ltd, 9600 Garsington Road, Oxford OX4 2DQ, UK
  Tel: +44 (0)1865 776868
Blackwell Publishing Inc., 350 Main Street, Malden, MA 02148-5020, USA
  Tel: +1 781 388 8250
Blackwell Publishing Asia Pty Ltd, 550 Swanston Street, Carlton, Victoria 3053, Australia
  Tel: +61 (0)3 8359 1011

First published 2008 by Blackwell Publishing Ltd

ISBN: 978-1-4051-6931-8

Library of Congress Cataloging-in-Publication Data
Richbell, David.
  Mediation of construction disputes / David Richbell.
      p.   cm.
  Includes index.
  ISBN-13: 978-1-4051-6931-8 (pbk. : alk. paper)
  ISBN-10: 1-4051-6931-1 (pbk. : alk. paper)
  1. Construction contracts–Great Britain.   2. Arbitration and award–Great Britain.   I. Title.
KD1641.Z9R525 2008
343.41′078624–dc22

                                                      2007032689

A catalogue record for this title is available from the British Library

Set in 10/12.5pt., Palatino by Aptara Inc., New Delhi, India

For further information on Blackwell Publishing, visit our website:
www.blackwellpublishing.com/construction

Contents

Acknowledgements   xi
Introduction   xiii

**Chapter 1 The Construction Industry Is Great at
Dispute Creation**   1

1.1	Contract	3
1.1.1	The 'no contract' scenario	3
1.1.2	The 'is there/is there not a contract' scenario	4
1.1.3	Incompatible contracts	5
1.1.4	All-risk contracts	5
1.1.5	Unrealistic performance criteria	6
1.2	Finance	7
1.2.1	A low-margin industry	7
1.2.2	Claim culture	7
1.2.3	The squeeze game	8
1.3	Culture	8
1.3.1	Adversarial	8
1.3.2	Fragmented	9
1.3.3	High failure rate	9
1.3.4	Complexity	9
1.3.5	Incurable optimism	10
1.4	External factors	10
1.4.1	Weather-sensitive	10
1.4.2	Consultants	11
1.4.3	Government legislation	11
1.5	Why people get into disputes	12
1.5.1	Communication	12
1.5.2	Personality	13
1.5.3	Interpretation	14
	Chapter 1 in a nutshell	15

Chapter 2 The Dispute Resolution Options   17

2.1	Consensual methods of resolving disputes	19
2.1.1	Negotiation	19
2.1.2	Conciliation	19
2.1.3	Mediation	19

2.1.4	Arb/Med and Adj/Med	20
2.1.5	Court settlement procedure	20
2.2	Resolving disputes through recommendation	21
2.2.1	Neutral fact-finding	21
2.2.2	Dispute review boards	21
2.2.3	Early neutral evaluation (also judicial appraisal)	21
2.2.4	Mediator recommendation	22
2.2.5	Conciliation	22
2.3	Imposed solutions to disputes	23
2.3.1	Med/Arb and Med/Adj	23
2.3.2	Adjudication	23
2.3.3	Ombudsman	23
2.3.4	Expert determination	24
2.3.5	Tribunals	24
2.3.6	Arbitration	24
2.3.7	Litigation	24
2.4	Why traditional methods fail the parties	25
2.4.1	Inherent injustice	25
2.4.2	Cost and time	25
2.4.3	Adjudication is not the 'saviour'	26
2.5	The better options for dispute resolution	27
2.5.1	Consensual processes	27
2.5.2	Partnering	27
2.5.3	Dispute avoidance	27
	Chapter 2 in a nutshell	28

Chapter 3 The Case for the Mediation of Construction Disputes 29

3.1	Better deals	29
3.2	Speed and economy	32
3.3	Flexibility in process and outcome	33
3.4	Finality of outcome	33
3.5	Mediator 'added value'	34
3.6	Getting off the treadmill	35
3.7	Ongoing relationships	36
3.8	Day in court	37
3.9	Commercial *v.* Legal	37
3.10	The arguments against mediation	38
3.10.1	It adds another layer of cost	38
3.10.2	It is too 'touchy-feely'	38
3.10.3	Mediation is non-binding so has no teeth	39
3.10.4	Mediation is all about compromise	39
3.10.5	Mediation is all talk, no commitment	40
	Chapter 3 in a nutshell	40

Contents

Chapter 4 Preparing for Mediation   **41**

4.1	Typical framework	41
4.2	Stages of mediation	41
4.3	Preparation by the mediator	44
4.4	Preparation by the parties	46
4.4.1	When to mediate	46
4.4.2	How long should the mediation take?	46
4.4.3	Who to choose as mediator	47
4.4.4	Mediator fees	49
4.4.5	Co-mediation	50
4.4.6	Assistant mediators	51
4.4.7	Conflicts of interest	51
4.4.8	Documents	52
4.4.9	Where to mediate	54
4.4.10	Who attends?	55
4.4.11	Authority	56
4.4.12	Who presents?	58
4.4.13	Dry run?	58
4.4.14	Pre-mediation meeting	59
4.4.15	Pre-mediation contact	61
4.4.16	The mediation agreement	62
4.4.17	Risk analysis	63
4.4.18	Anticipate the settlement	65
	Chapter 4 in a nutshell	66

Chapter 5 Presenting at the Mediation   **67**

5.1	Arrival	67
5.2	Pre-meeting	68
5.3	Initial joint meeting	69
5.4	Using the joint meeting	71
5.5	Exploring stage	73
5.6	Giving and receiving information	75
5.7	Idle time	76
5.8	Others' shoes	77
5.9	Non-financials	78
	Chapter 5 in a nutshell	79

Chapter 6 Negotiating at the Mediation   **81**

6.1	Negotiation zones	81
6.2	Reviewing	82
6.3	Bottom lines	83
6.4	Negotiation strategy	84

6.5	Incompatible styles	84
6.6	First offers	86
6.7	Offer logic	86
6.8	Getting into deadlock	87
6.9	Pain–pain	88
6.10	Keeping options open	88
6.11	Non-financials	90
6.12	And finally	90
	Chapter 6 in a nutshell	90

Chapter 7	**Concluding the Mediation**	**91**
7.1	Finality	91
7.2	Deals with dignity	91
7.3	Part-deals	92
7.4	No deal	93
7.5	Writing the settlement agreement	95
7.6	What can go wrong?	96
7.7	Cooling-off period	97
7.8	Mediator recommendation	98
7.9	Mediator liability	99
	Chapter 7 in a nutshell	99

Chapter 8	**Roles in Mediation (Who Does What?)**	**101**
8.1	Client	101
8.2	Legal advisor	102
8.3	Counsel	103
8.4	Experts	105
8.5	Consultants	106
8.6	Support staff	107
8.7	Mediator	108
8.8	Assistant	109
8.9	Co-mediator	109
	Chapter 8 in a nutshell	110

Chapter 9	**Avoiding Disputes in the Construction Industry**	**113**
9.1	The positive side of conflict	113
9.2	Understanding – valuing – cooperating	113
9.3	Creating a culture that is positive	114
9.4	Twelve rules and challenges	115

9.4.1	Establish clear, simple and constant lines of communication	115
9.4.2	Establish clear roles, responsibilities, accountabilities and systems	115
9.4.3	Practise (and therefore model) openness/ transparency	116
9.4.4	Build trust from the start; cooperate rather than confront	117
9.4.5	Acknowledge problems, don't bury them	117
9.4.6	Treat mistakes as learning points, not blame-makers	117
9.4.7	Get the 'headline' agreed	118
9.4.8	Listen, and show that you have heard	118
9.4.9	Establish what parties need, rather than what they claim	119
9.4.10	Involve a neutral early when disagreements are unresolved	119
9.4.11	Re-evaluate agreements and headlines in the light of resolution	120
9.4.12	Re-commit to the relationship/contract	120
9.5	Partnering	121
	Chapter 9 in a nutshell	122

Chapter 10 The Mediation Landscape		**123**
10.1	Deal mediation	123
10.1.1	What is it?	123
10.1.2	What does the deal mediator do?	124
10.1.3	Difference between a deal mediator and dispute mediator	125
10.2	Project mediation	126
10.3	Dispute mediation	127
10.4	Facilitation	128
10.4.1	Preparing for the facilitation	128
10.4.2	Agenda	129
10.4.3	Structure of the day	129
10.4.4	Open space	129
10.4.5	Role of mediator as facilitator	130
10.5	Consensus-building	130
10.6	Bespoke mediation processes	131
10.6.1	Construction Conciliation Group (CCG)	131
10.6.2	RICS Neighbour Dispute Service	132
10.7	Tiered resolution	133
	Chapter 10 in a nutshell	133

Chapter 11 Conclusion – How to Win at Mediation   **135**

11.1	Prepare well	135
11.2	Chose the right mediator	135
11.3	Get the best out of the opening joint session	136
11.4	Cooperate	136
11.5	Have a drink!	137
11.6	And remember ...	137

Appendices   **139**

Summary of relevant law	141
Risk analysis checklist	148
Decision tree	150
Pre-mediation checklist	153
Typical mediation agreement	155
Typical settlement agreement	157
Model Settlement Agreement and Tomlin Order	158
Mediation providers	162

Acknowledgements

Inevitably there are several people whose influence, support, comment and contribution made this book a pleasure to write. The danger about naming them is that some may be left out – but name them I will because they should have acknowledgement ... and if your name should be amongst them, and is not, please accept my apologies and let me know so that the second edition may be corrected!

From the construction industry David Cornes (Lawyer Mediator), David Canning (Highpoint Rendell), Simon Toulson (Nabarro Nathanson) and Philip Naughton QC (Mediator and Arbitrator).

Jane Gunn (my training partner in MATA) read every chapter as it was produced and made some very important suggestions in a tactful way. As with Chris, my wife, and Jo, my 'secretary' (although 'totally runs the office and gets David to the right place at the right time with the right papers' would be a more correct title), Jane has sustained me with encouragement and good humour throughout. Jo did the graphics and Chris read through and corrected the finished article (enough to send anyone to sleep on a regular basis).

David Miles (Construction Law Partner at Glovers and to whom this book is dedicated) and Tina Monberg (Danish Mediator and Trainer) read and commented on specific chapters, and John Clarke, Roger Tabakin, Andrew Acland, Chris Cox, CEDR and Pepperdine University all gave permission for me to include their work as diagrams or appendices.

Finally Julia Burden, the Commissioning Editor at Wiley-Blackwell, had the courage to agree to the book and said nice things at critical moments.

Thank you all – not least because you gave time, support and help freely.

Introduction

I have spent all my working life in the construction industry. I love it. Even though the last ten years or more have been spent resolving disputes, and so working away from the daily involvement, whenever I attend a gathering of construction professionals I immediately feel at home and in great company. There always seems to be an air of optimism and good humour despite the daily battles and the difficulties of earning a living and the challenges of the latest health and safety regulation. The adversarial climate does not seem to prevent people having a drink together . . . and then returning to the battle the next day. And the building or structure that results is usually something that building's owner, consultants and contractors can be proud of. It is a very positive, and generally creative, industry.

However, it has to be said that one of the things the construction industry is most creative about is disputes. And it is also, in my view, very bad at resolving them. Invariably the outcome is a win–lose result where the winner takes all and the loser pays. But life is not like that. There is always more than one version of the truth but the traditional ways of resolution (arbitration, litigation, adjudication) cannot take that into account. A decision must be made, someone must win, and therefore someone must lose. This has nothing to do with justice, and it is amazing to me that the construction industry has been content with such an unjust resolution process and not sought a fairer and more common-sense-based process. It is even more amazing to me that mediation, now a tried and tested process that is based upon common sense and inevitably achieves outcomes that are just – or, at least, prudent – has been largely ignored because adjudication has become the favoured process. It seems to me that this situation is now changing and mediation is becoming increasingly used in the resolution of construction disputes. And the more it is being used, the greater becomes its reputation as a sensible dispute resolution process.

This book is written for users of mediation, whether they be parties, advisors or experts. It should also be of help to commercial mediators who have no specialism in construction and who wish to be alert to the peculiar nature of the construction world. It is not, however, a book to train mediators in the core skills and techniques of mediation. If that is what is needed, look elsewhere … and then read this as a follow-up. The underlying reason for this book is to encourage confidence in the mediation process and to ensure that those who use mediation to resolve their disputes do so effectively and so are able to maximise the opportunities that mediation offers.

Although mediation has been around in the UK for nearly 20 years, it is a great sadness to most commercial mediators that it is rarely used well by participants (parties and their advisors) and the opportunities that it offers to achieve creative solutions and restore previously successful relationships are rarely grasped. Much of this is due to ignorance, still. But much of it is also due to the fact that lawyers are the gatekeepers of mediation – they advise their clients about it, choose the time and select the mediator – and so they invariably take their legal approach into the mediation arena. Not all do so, but most do. The mediation forum becomes another place in which to continue the adversarial litigation process; and this is reinforced by the increasing number of barristers who attend mediations, who naturally assume that they should lead and speak on behalf of their clients and ensure that their legal opinions are seen to be correct, even though the other side's counsel argues the opposite. Again, not all do so, but most do.

It is the parties who have most to gain from mediation (although a settlement in mediation invariably provides the lawyer with a satisfied client, which must be good for potential future business). The party gains not only in obtaining sensible outcomes but more so because of avoiding the wasted management time involved in pursuing or defending an action. No pre-estimate of management time is ever accurate and the disruption to normal business can be immense. Key people become unavailable when most needed, resources and money are diverted to activities other than wealth-generation and genuine people are made to look silly when cross-examined by counsel. Mediation offers the chance to avoid all that at an early stage and provides a quick and cost-effective resolution of parties' disputes.

It is also amazing to me that lawyer mediators tend to be preferred to non-lawyer mediators. That may well be because lawyers are doing the choosing and they prefer a mediator who comes from their environment and speaks their language – and perhaps may be malleable towards their version of the truth. However, mediations are never settled on legal argument for they always become commercial negotiations; and who better to facilitate effective commercial negotiations than a mediator who comes with a business, rather than legal, background? Mediation is, after all, just an assisted negotiation. Better to have someone who is skilled in negotiation – and I mean business negotiation, not legal argument – to help parties achieve the best and lasting deal.

The first three chapters of this book are devoted to making the case for mediation in an industry that is traditionally adversarial and seeks an independent third party to resolve disputes that have not been resolved by the parties unaided. It draws upon the survey carried out by the Construction Industry Forum (CIF) of Ireland which identified that 2% of its members' turnover was being spent on managing disputes. This survey led to the CIF commissioning a dispute avoidance roadshow to help its members channel more resources into dispute avoidance – and therefore into wealth-generation – rather than into the extraordinary (or should that be 'ordinary'?) wastage involved in creating and fighting disputes. The book then looks at the mediation process itself and the roles played by those involved. The second, and probably most important, section (Chapters 4 to 8) looks at how to use the process to best advantage and looks at the variations of the mediation process that are available. The final section (Chapters 9 to 11) looks at how to avoid using a mediator in the first place and deals with dispute avoidance, drawing on the CIF experience. The appendix then provides samples of standard forms, as well as a summary of relevant law.

There are several key statements that recur throughout this book:

- People see the same events/facts through different eyes. It doesn't have to mean that they are any more right or wrong than one another, just different.
- As a result, people believe their version of the (same) truth.
- Mediation allows parties to tell their story (truly to have their day in court).
- The Mediator is there to give the parties the best shot at doing a deal.

There are probably many other repeated statements but these four are key to the theme of this book and underlie my belief that forms of dispute resolution other than mediation are inherently unjust.

The construction industry is in my blood. Even in the seventeenth century a Richbell was building houses in what is now the London Borough of Camden and 'Richbell Court' gives evidence to this. For over 40 years I lived the business, initially with contractors and then ultimately having my own quantity surveying and project management practice. Sunday walks with the children were opportunities to point out chimney pot designs and to admire craftsmanship in stone, wood and brick. Even now, over ten years on from giving it up to become a commercial mediator, my landmarks are buildings and my most sociable drinks are with contractors. When I trained as a commercial mediator in 1991 I knew immediately that this was by far the best and most sensible way of resolving disputes, so much so that I ceased my QS practice in 1996, initially to become a director of CEDR (Centre for Effective Dispute Resolution) and then, from 2001, to practise on my own. Throughout that time I have trained mediators and also lawyers in how best to use mediation and now, as MATA (Mediation and Training Alternatives) in partnership with the Chartered Institute of Arbitrators, in delivering our Mediator Training Course. We also run an annual international Advanced Mediator Training Course, usually in Italy where the weather and wine add to the learning experience.

This is my attempt to deliver to the industry that I love a book that will help restore common sense to its disagreements and so allow it to become an even more human and respected industry than the one that I left.

CHAPTER ONE

The Construction Industry Is Great at Dispute Creation

This chapter looks at the reasons for the construction industry being notorious for creating disputes – a reputation that is worldwide – and ends with an examination of why people get into dispute in the first place. The construction industry is a great industry for creating disputes, and not just in the UK. As a quantity surveyor, first with contractors and then in private practice, I grew up with disputes, probably caused some of them, certainly fanned some of the flames and delighted in most of the battles. It was a wonderful training ground, both in the causes of disputes and in the appalling way in which they were resolved. Little has changed since I gave it all up in 1996 to become a full-time mediator and trainer. And once a year for most of the past ten I have spent a day with 50 or more graduate chartered surveyors, talking about collaborative negotiation and dispute avoidance. Few have really wanted to know. Their presumption is almost always that contractors/sub-contractors are out to screw them (or rather, their clients) – that's how they make their profit – and so being cooperative is a sign of weakness. Anyway, they are all in their twenties: energetic, competitive, most of them self-assured. They like the fight; they like to *win*, no matter what the cost.

Of course, the construction industry has tried various ways to avoid the endless disputes. Even in my quantity surveying prime, partnering was our saviour, though no one really knew what it meant – and they still don't, it seems. I recently mediated a dispute between a local authority and a contractor who had been partnering for several years. But it wasn't really partnering: no real training about working cooperatively and no defined structure for resolving differences. Yes, there was a 30-page partnering agreement drawn up by lawyers, but no one really knew what it meant. People tried a bit harder to be nice to each other for a period, then something went wrong and partnering was forgotten, and the old ways returned. It is almost as if there is comfort in the familiar, no matter how inefficient and ineffective it is.

Other routes, such as design and build, were intended to put all the responsibility onto the contractor and so render the building owner free from claims. 'Fixed price lump sum' was supposed to be just that, but how many of us smiled when the Football Association used that term regarding the delays on the Wembley contract? 'Fixed price ... we won't be paying anything more,' they said. 'Oh yeah!' was the universal reaction; that's not how it works.

There is not an area of construction activity that is protected from this dispute-orientated environment.

Case study

In 2005 there was a report on a dispute involving a couple who had an extension built to their house – quite a large one, and built by the most reputable builder in the area. He was paid up to the last £5,000, but that last amount was withheld because the owners were dissatisfied with some, quite minor, work. The contractor pulled off site, refusing to complete the work until the monies due to him were paid. No money came, the contractor instructed solicitors and, some time later, after the case was heard and appealed, the building owners were faced with having to sell their house to pay the £250,000 total legal costs. The appeal judge was critical of the lawyers for allowing such cost to accumulate, but – to a greater or lesser extent – this is not an uncommon situation.

I recently mediated in the IDRC building[1] and their reception was filled with a recent delivery of around 30 bankers' boxes of documents – one side's papers for an arbitration that was destined to last several months. Obviously a big case (well, I hope so!). Good for the IDRC who rented out the rooms and facilities. Good for the arbitrators (I think there were three). Good for the lawyers. Bad for the parties and bad for the reputation of all involved. At the end of the day the excessive time spent in resolving a dispute, and the consequential cost, is not something to brag about. Many small (that is, up to £250,000) disputes that I mediate are in the awful situation of the costs exceeding the claim, so the dispute is no longer about settling a claim but about who pays the costs.

Not only does the construction industry have no regard to the size of the project for breeding disputes, but its disputes also affect the whole spectrum of its activities. Client/contractor, contractor/sub-contractor,

---

[1] International Dispute Resolution Centre, 70 Fleet Street, London.

client/consultant and any permutation of those that you care to imagine – even better, *all* of them in one dispute: final account, extensions in time, loss and expense, professional negligence, performance, defects and so on, and so on. There is no limit to the possibilities. All because the construction industry loves a fight and hates to change.

Though endemic, disputes are bad for the industry. The CIF[2] survey of the Irish construction industry carried out in 2006 suggests that 2% of turnover is spent on managing disputes. In a 3% margin industry, that must be bad. Even if the Irish experience is more extreme than that in the UK, the amount of lost wealth caused by disputes is huge.

Disputes not only waste money, and therefore drain profits. They sour and even destroy relationships. They can be stressful, and physically and psychologically draining. They take attention and energy away from the project, and the focus is no longer on successful completion but on the impending fight. Disputes limit and often distort communication, raise suspicion and threaten financial instability. They make other problems more difficult to resolve. Only negatives result from unresolved disputes.

It is apparent that the reasons for disputes being so much a part of construction life fall into four categories:

- Contract (Is there one? Is it enforceable? Where are the 'get-outs'?)
- Finance (low margin, claim culture, credit-based)
- Culture (adversarial, fragmented, incurably optimistic)
- External factors (consultants, weather, legislation)

1.1 Contract

1.1.1 The 'no contract' scenario

It is amazing to me how many disputes I mediate where no contract exists at all. It is usually the parties' intention to draw up a contract, but whether it be because the parties are optimistic that there will be no problem, or because they feel they have a good enough relationship to be able to weather any difficulties, is not clear. One thing that does appear to be common in these disputes, though, is the eternal optimism of the construction industry that this job will be the perfect one: the one that has no delays, is designed to work perfectly and programmed to

---

[2] Construction Industry Federation, Dublin.

be built in weather that is neither wet nor too hot. As if anyone had any experience of a project that justifies such optimism!

Of course, when problems do occur there is invariably disagreement over what the contract would have said if it had existed or, if one was clearly implied, disagreement over the true meaning of what it would have said. Because problems have arisen that have not been sorted (and of course I only get to mediate those that have not been sorted), it usually means that relationships have deteriorated and extreme positions have already been taken. Hurt pride and/or broken trust tend to underlie most of these disputes and initial good intentions are of no importance when difficulties become disputes. Most people revert to positions and become defensive, whereas when they originally set up the project they had every intention of working cooperatively because their trust was absolute. Sometimes it almost appears to be an insult to suggest that the relationship is formalised into a contract. It is almost like saying 'Don't you trust me?'. Yet, regrettably, experience shows that having a contract to fall back on that is clear about roles, about who pays for what and how disputes may be resolved, is a wise move. I must remember to do so on my next domestic building project!

Unfortunately, even when contracts do exist, they may not be absolutely clear and free from dispute.

1.1.2   The 'is there/is there not a contract' scenario

Most construction projects that do have a contract use a standard form. In theory, standard contracts remove most of the uncertainty and, as a result, everyone understands their meaning and applies their conditions consistently and universally. In theory. And there must be many projects where this is so – after all, I only get involved in the ones that go wrong. But the ones that do reach me almost always revolve around the legal arguments of

- whether or not the contract existed in the first place, or
- the contract is at large because parties did not fulfil their obligations.

Lawyers tend to love these arguments. After all, this is what they are trained for: interpreting contracts, knowing the law that supports their arguments and building the best case they can to support their client. Many times I have heard arguments run that fill me with admiration for their creativity and conviction. Ever optimistic, the lawyers have devised and run even the most obscure and unlikely arguments to prove that, assuming there was a contract ('which we may not accept'), their interpretation 'in these particular circumstances' is correct, irrespective of the commonly accepted interpretation that it defies. That must be one of the few joys of being a litigator – fighting lost causes in the hope that they are not lost! Of course, this is the exception. The vast majority of the cases I mediate have lawyers who have given a realistic assessment of the party's chances of succeeding in court, and that assessment has been accepted by the party. If the chances are poor, they rarely try to bluff it out in the extreme way I have suggested. But some do, and whilst it spices up the day, it inevitably makes settlement more difficult.

1.1.3   Incompatible contracts

Whilst the parties may have entered into a clear and sensible main contract, there will be others with sub-contractors and specialists. The intention will always be for contracts to be 'back-to-back' so that the conditions of the main contract are replicated in the sub-contracts. I expect most are, but I frequently mediate construction cases where they are not. The intention may be there, but whether it be because of time pressures or particular conditions that the sub-contractor/specialist may have or unresolved differences that rumble on even whilst the work is being carried out, the target of all contracts being back-to-back is often missed. This is particularly common with consequential losses and with liquidated and ascertained damages. It is very difficult to make a small but key specialist sub-contractor responsible for full LADs, yet sometimes their work can be on the critical path and so delays may have a significant effect on the progress of the project as a whole.

1.1.4   All-risk contracts

One-sided contracts inevitably fail, and that inevitably leads to disputes. There has been a trend over many years for consultants to remove any client risk and so attempt to put it all onto the contractor. Not only does this breed a feeling of injustice, but when things go wrong and the contractor loses, there will inevitably be an attempt to shed the risk. Standard contracts attempt to spread the risk – some with the building owner (changes, delayed access, etc.), some with the contractor (delays, workmanship, etc.) and some shared (inclement weather, acts of God, etc.). That is for a reason, and when the risk (even, in one case I mediated, for inclement weather on a major earth-moving contract) is shifted onto one party, the balance is changed and the relationships affected. The Wembley Stadium contract is an example where the acceptance of risk by the contractor caused such a huge loss that a deal had to be negotiated before the Stadium was released to the Football Association – despite the FA saying publicly, 'We have a fixed price contract and will not be paying a penny more'.

1.1.5 Unrealistic performance criteria

Construction is not a precise science. It is subject to weather, to human preferences and skill (or lack of skill), and to natural or manufactured materials that have their own characteristics. Yet consultants frequently produce specifications and designs that set unrealistic criteria and assume perfection, and then blame the contractor when their aspirations are not met. There have been many expert battles over what is and isn't a reasonable standard. And alongside that argument is the question of what is the reasonable remedy when the finished work fails to meet the consultant's or, worse, the building owner's expectation: rip it out and start again (and suffer the consequential time penalties), or repair (and the owner suffer potential increased maintenance costs or inferior finish)?

Case study

Concrete to an airport runway was shelling at the construction/movement joints and the airport authority was concerned about concrete fragments being drawn up into the aircraft engines. Experts took core samples of the concrete slabs and found some areas of concrete to be of inferior strength to that specified. The contractor was confident that a careful repair would sort it out (carried out at night whilst the airport was closed) although the new concrete would obviously be seen to be patched. The airport authority required total replacement of the affected concrete slabs (which would involve closing down the runway and restricting traffic as a result – and a consequential claim against the contractor). It was also not very realistic because adjoining slabs were likely to be damaged when breaking up the concrete. Another option was to overlay the concrete with heavy-duty tarmacadam, which could be done overnight and would avoid the breaking up of the allegedly

---

> defective concrete, but it had a lower life expectancy and so main-
> tenance costs for the airport would be higher. A perfect result was
> never a possibility. The final result is still not known to me as the
> mediation was over other matters.

1.2 Finance

1.2.1 A low-margin industry

I do not know why the UK construction industry has created the tradition of being such a low-margin industry. It doesn't seem to be so in
most other countries, yet in the UK tenders are won on a 3% margin
or less. That leaves very little room for error and so inevitably creates
a culture in which claims become the rule. How else is a contractor to
make money? It is no surprise that many contractors nowadays only negotiate contracts or look to partnering arrangements where a fair return
is a more recognised principle. But with competitive tendering still the
normal process, it becomes a money-management exercise where contractors have to work the cash flow to make a profit. How the industry
worked itself into such a silly situation I know not, but it is a tradition
that is very difficult to break from. The cheapest price is still likely to
get the order.

1.2.2 Claim culture

Unfortunately, because it is a low-margin industry and subject to the
uncertainty of the British weather, claims are an inevitable part of the
culture. It is still said that some contractors start their claims before they
get on site. Every document is scrutinised in detail and every drawing
analysed with the intention of finding inconsistencies which can be ex-
ploited. Letters are written and programmes prepared which can give
the maximum advantage to the contractor, and the maximum possibility for financial gain. What a shame that so much resource is devoted
to such negative activity. It puts everyone involved in the project on
the defensive, energies are channelled into the wrong activity and focus
on completing the project on time and to budget is diverted. And to
what effect? I had a mediation recently where the defendant (a building owner) said, 'Everyone knows that claims are three times as much
as they should be so tell them to reduce their claim by two-thirds and

then we will start negotiation.' A somewhat cynical approach to negotiating a deal, but experience does show that it might be realistic. Such
is the culture, fuelled by claim specialists, and it makes genuine claims
very difficult to settle. How much better it would be for members of the
construction industry in the UK to achieve a fair price for the work at
the beginning and so channel their creative resources into completing a
high-quality job on time and within budget.

1.2.3   The squeeze game

The Housing Grants and Regeneration Act 1996 attempted to deal with
contractors squeezing sub-contractors by withholding money and pres-
suring the small firms into lower prices or carrying out additional work
unpaid. There is no doubt that many small firms are still in existence
because of the Act and so it has worked at that level. However, it rarely
brings finality and it is not an unusual experience for squeezing to happen all the way down the line: employer to contractor to sub-contractor
and beyond.

1.3   *Culture*

1.3.1   Adversarial

The construction industry seems to love confrontation. The method of
obtaining the work and the process of agreeing the final account are
founded on fighting it out. Competitive tendering pitches contractors
against each other, the theory being that they can bring special expertise
to the construction process that will give them an edge and so undercut
their rivals. In reality, competitive tenders are won as much on estima-
tors' mistakes, or over-optimism, as on expertise or unique resources.
Even reputation comes second to the lowest tender, and although the
consultants usually say in their tender conditions that their client will
not commit to accepting the lowest, or any, tender, it is unusual for
building owners to pay significantly more than the lowest price just
because of another contractor's reputation. It begs the question of why
they invite the lowest-priced contractor in the first place (and put them
to the significant cost of preparing a detailed tender) if their tender was
not going to be accepted. In theory, negotiating tenders removes some
of the confrontation and partnering removes it totally. Unfortunately,

that is not the case in many contracts as adjudicators, arbitrators and, increasingly, mediators will happily testify.

1.3.2 Fragmented

Fifty years ago, contractors employed a comprehensive workforce that offered all trades and skills in-house. Over those 50 years and initially motivated by the increasing cost of employing people compared with the attraction of using cheaper self-employed people, the industry has become fragmented. Even in 1995, 95% of construction businesses had eight or fewer employees. Few, if any, contractors have more than just a small core of staff, having become little more than managers of the construction process. Craftsmen, even labourer gangs, have become sub-contractors and sell their trade wherever it is needed, usually to the highest bidder. It creates the opposite of a seamless process and often progress on projects is controlled by the availability (or lack of availability) of relevant labour at the critical times. Not only is a seamless process very difficult, but the passing of risk down the construction line is almost impossible. The more fragmented the workforce, the smaller their substance and the lower the chances of passing risk. In the end, the main contractor has to take the risk whilst having very little control over those who create that risk.

1.3.3 High failure rate

The fragmentation of the industry and its adversarial culture cause numerous business failures each year. In 2005 the construction industry, the sixth largest industry in the country, accounted for 5% of all company failures in the UK. It is a low-margin high-risk industry and the chances of failure are high. And the fight for survival increases the likelihood of disputes, making the need for mediation even greater. After all, a company fighting for survival will have little financial resource to fund litigation or arbitration. Mediation is a quick and cost-effective way of resolving even the most complex disputes.

1.3.4 Complexity

Construction is a complex process, not only involving many different, and often small, sub-contractors but also a multitude of consultants, each having their own needs and demands. Even the smaller projects are likely to have designers (building and structural) and the ever more powerful custodians of the health and safety requirements, plus the local authority planning and Building Regulations officers. Multiply this for larger projects to include architects, quantity surveyors, project managers, employer's representative and so on, plus an often demanding building owner – this is a lot of people to please and it often results in incompatible demands.

And then there is the financial complexity. Those projects with bills of quantities (and there still are some!) have every item of work measured in infinite detail which is then priced – in infinite detail. Quantity surveyors can spend time calculating, and then arguing over, rates for various items of work that is quite disproportionate to their value. Yet, ultimately, the only thing that really matters is the sum at the end – does it cover the cost (and give a return)?

1.3.5 Incurable optimism

One of the things I admire about the UK construction industry is – despite all the problems, the low margins and the impossible task of pleasing everyone all the time – the incurable optimism of its members. An American arbitrator of great age once commented that the British culture after the Second World War was one of paranoia about unemployment and so the construction industry's main focus was to stretch the job to keep people in work. I hardly think that this could still be true but it may be that the optimism is partly from having another year with a job! Most people in the construction industry have a 'can do' approach and take pleasure in creating what someone else has designed and specified and commissioned. Of course, bearing in mind the failure rate and the number of disputes, this optimism could be seen to be misplaced. The assumption that this is the job where everything goes right should have been tempered long ago by the experience that no such job ever existed. It is how you cope with the problems when they do arise that matters most.

1.4 External factors

1.4.1 Weather-sensitive

The construction industry is intensely weather-sensitive. Ideally all work in the ground is done in the summer, the roof is on and finished before the rain starts, and heating is commissioned in the cold weather. Sod's law dictates that, even if the foundations are excavated and cast in the summer, it will be the wettest summer in living memory and when the roof is about to be covered the strongest winds since records began will strike, probably blowing off the temporary covering just in time for another deluge to soak all the floors and walls now uncovered. Even when the weather is kind and the sun comes, out it will probably be a heatwave and cause excessive shrinkage to all the finishings. And all this in a low-margin industry where most of the work is outside!

1.4.2   Consultants

Most architects and engineers long for projects where they leave a personal signature. Quantity surveyors can only take pride in their documentation (and protecting the posteriors of their consultant colleagues), but designers have the ability, and often the need, to leave their signature. However, few consultants can ignore the spectre of professional negligence claims, and this seems to be a dominant and driving influence on the way that they conduct their business. It must be very difficult to maintain professional standards and independence in such circumstances. It is a recipe for caution that stifles innovation and creativity. Few clients will promise not to sue their architect or engineer if their innovatory design does not work, or has residual problems. In a blame-finding culture, most consultants will avoid risk wherever they can and transfer responsibility to others (usually the contractor!). And with professional indemnity insurance premiums as high as they are, who can blame them?

1.4.3   Government legislation

Last under the heading of External Factors is the government: producing more and more legislation that affects the construction industry, adopting more and more regulation from Europe, all imposing more and more burden on the contractor and often on the consultants. Much of it causes yet more restriction on the materials that can be used, which in turn places limits on the design options. It may be safer for the workers, and more environmentally friendly, but it also adds further complication to an already complicated process. The ground for misunderstanding and disagreement has never been more fertile!

1.5   Why people get into disputes

All of this would be avoided, or be much simpler, if it were not for the people. People are complex, different, difficult, have preferences and opinions. They are ideal dispute-creators.

There has been a lot of research into personality types and how different people react differently to conflict. For our purposes I summarise them into three headings:

- Communication
- Personality
- Interpretation

1.5.1   Communication

Bad communication is almost always the start of conflict and, if not the start, it is the cause of conflict escalating into disputes. The main causes are as follows:

- The form of communication. The written word is the worst form of communication. So often the meaning of the writer is interpreted differently, leading to misunderstanding and conflict. In an age of email and text-messaging, this means that communication is potentially at its worst. The spoken word (by telephone) is marginally better because the tone, the pace, the emphasis all colour the words, so the chances of misunderstandings are reduced, and if they do arise, clarification can be obtained immediately and conflict avoided. Even then, we are only about one-third effective in our communication. Body posture, gestures, contact and personal space are all part of human communication. We understand so much more of the message being imparted by reading a person's body language. This can be demonstrated by observing people speaking a foreign language, or viewing people in conversation who are out of earshot. So much of their conversation can be interpreted without hearing the words.
- Attitude. We are generally bad listeners and want instead to have our say: determined to get our view across, to convince the other person that our position is right. Most of us show little interest in the other's point of view, give minimal response to what the other person has said and probably follow up with a change of subject or with our own

similar (and probably better) experience. All this results in a rather one-sided conversation.

The key to effective communication is to establish clear, simple and constant lines from the start, to recognise that face-to-face meetings are twice as effective as the telephone and ten times as effective as the written word, and to be prepared to really listen.

1.5.2 Personality

We are all different and we react to conflict or difficult situations in different ways, depending on our personality. A lot of research has been done on people's attitude to conflict. Thomas and Kilmann[3] developed (in 1974, so it is well tried and tested) a model for explaining different approaches to handling conflict. The premise is that each of us has a preferred style of handling conflict – and that all of us can learn to work outside our preferred style. This is a brief run through the five personality styles.

Competing

Assertive and uncooperative; an individual pursues their own concerns at the other person's expense. Useful when a quick decision is needed. Not good for team-building.

Avoiding

Unassertive and uncooperative; an individual does not immediately pursue their own needs or those of the other person. They do not tackle the conflict. Useful if they are not the best person to solve the problem, or when time is needed to change the situation. Not good when decisions are needed, and can be frustrating to others.

Collaborating

Assertive and cooperative; explores positions to try to find a solution that fully meets the interests of both parties. Requires trust and time. Useful because this usually provides the best solution to a problem. Not good when immediate decisions are needed, and can be frustrating for people in other styles.

---

[3] Thomas-Kilmann Conflict Mode Instrument (TKI).

Accommodating

Unassertive and cooperative; an individual accepts the other's position without pursuing their own. Useful when they are wrong, and when pursuing the issue will cause more damage to the bigger picture. Not good if own needs go unfulfilled.

Compromising

Assertive and cooperative; the middle ground. Usually means that each party gets something and so has some satisfaction. Not good because the issues are rarely explored in any depth and so the solutions are rarely the best.

We each have our preferred style and like operating in this comfort zone but problems often arise when the styles within a group clash. People who are competitive get impatient with avoiders and collaborators, and vice versa. Recognising and adapting to another person's preference will help with creating an effective level of communication.

1.5.3 Interpretation

People enter into relationships with their own expectations, presumptions and interpretations. Because people see the same events/facts through different eyes, it does not always mean they are any more right or wrong: just different. Conflict arises when interpretations differ and these differences need exploring so that the reasons can be understood and valued. The trouble is that most people are either not inclined, or do not have the time, or perhaps just lack the sensitivity, to be an enquirer.

There is nothing worse than having no clear idea where we are going. This problem often starts with the initial goals of a relationship/project. People have a different understanding of the required outcome, or just visualise an outcome that suits their needs rather than other's. Getting the 'headline' right (and agreed) at the outset is vital so that everyone has a common vision and they are heading in the same direction. This applies as much to a simple business negotiation as it does to a major project. The headline is the touch-point for all future decisions.

Another problem area, when people do communicate, is whom it should be with. It is important to establish clear roles and responsibilities from the start so that everyone knows what is expected of them and to whom they report. Simplicity is everything. Decisions should be made at the lowest possible level of responsibility. People's strengths should be used to the full, their weaknesses recognised and compensated (e.g. meetings can be a waste of time if they are not chaired well – choosing the chair is more important than allowing the senior person present to assume the role).

All of this comes down to effective communication. Time spent at the start of a project getting the structure and lines of responsibility agreed can overcome most of the problems that lead to disputes. The construction industry is great at creating disputes, but it could be different.

Chapter 1 in a nutshell

- The adversarial, low-margin, claim-based culture of the majority of the UK construction industry breeds disputes.
- Disputes waste time that could be spent on wealth-generation. They divert energy and move the focus from completing a high-quality project on time and on budget to defending/pursuing a claim.
- Poor communication is at the heart of most disputes. Face-to-face communication is ten times more effective than the written word.
- People react to conflict differently. Understanding and adapting to a person's style can help effective communication.
- Most people are bad listeners. We prefer to tell our story rather than listen to theirs.

CHAPTER TWO
The Dispute Resolution Options

Traditionally the construction industry has resorted to litigation or arbitration to achieve a resolution of its disputes. Then, following the Latham Report in 1994, adjudication became a statutory right under the Housing Grants and Regeneration Act of 1996. In 2006 the Royal Institution of Chartered Surveyors alone did over 2,000 adjudications. But all of these methods involve an independent person (or persons) handing a decision to the disputing parties, which results in an outright winner and therefore an outright loser. That is their weakness, and the result, dare I say, has little to do with justice.

The purpose of this chapter is to briefly outline the most common methods of resolving disputes in the UK as a background for a more detailed exploration of mediation in subsequent chapters. ADR ('alternative dispute resolution' but sometimes interpreted as 'appropriate dispute resolution' now that mediation is mainstream in the UK) covers all that follows except arbitration and litigation (which were, at the time of the Woolf Reforms in 1999, considered to be the established forms of dispute resolution in the UK). Overseas, where arbitration is not so established, ADR refers to everything other than the courts, and so includes arbitration.

This chapter divides the dispute resolution spectrum (Figure 2.1) into three main groups:

- Consensual – the parties in dispute agree a solution, sometimes with help from an independent party.
- Recommended – a third party investigates and recommends a solution which the parties do not have to accept.
- Imposed – a third party makes a decision that is binding on the parties.

The boundaries are not always clear and some processes on the spectrum overlap.

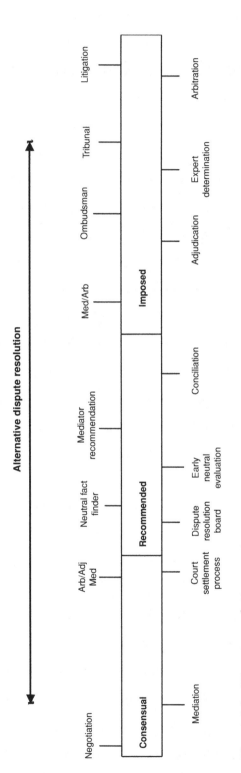

Figure 2.1 Dispute resolution spectrum.

2.1   Consensual methods of resolving disputes

2.1.1   Negotiation

Negotiation is the best and most efficient method of resolving disputes. It involves fewer people, takes less time and usually results in strengthened relationships. It is something we all do all the time, even if we do not recognise it. The best negotiations involve effective communication and a willingness to compromise, and result in the negotiating parties' needs being met. If one or more of these is not achieved then the negotiations usually deadlock. Negotiations can be public or private, depending on their nature and the wishes of the parties.

2.1.2   Conciliation

Conciliation is a term that is often confused with mediation. Unfortunately, in the UK it has several different meanings depending on the sector of dispute. As a consensual process it is often used in sectors such as the health service and some employment disputes as an informal stage in the complaints process. If negotiations between the complainant and the doctor/authority/employer do not achieve a settlement, a third party is brought into the discussions to facilitate a settlement. Conciliation in this context is usually a private process. However, conciliation in the construction industry is a more formal and structured process and occurs later in the spectrum.

2.1.3   Mediation

Mediation is a more structured form of assisted negotiation. It is a voluntary (unless required by contract), flexible process within a framework of joint and private meetings where the mediator helps the parties clarify the key issues and construct their own settlement. Most mediations last a day, and the vast majority settle on that day. Since the Civil Procedure Rules (CPR) were introduced in 1999 (the Woolf Reforms), mediation has become a mainstream process in the UK.

Many courts now have a fixed-fee, time-limited mediation scheme and the courts often give a strong nudge to the parties to use this voluntary process. Indeed, in what it considers to be appropriate cases the commercial court virtually insists on parties taking their case to mediation before trial. Since the Woolf Reforms the court is seen as 'the last resort' for resolving disputes, although in a ground-breaking case, *Hurst* v. *Leeming*,[1] the court did recognise that there are circumstances in which a party is justified in rejecting mediation.

Besides returning the control of their dispute to the parties, one of the main advantages of mediation is that the settlement can include whatever the parties decide (provided that it is legal). There is no outright winner or loser; all parties have to say 'Yes' for the deal to be done.

There are variations of mediation, some of which are mentioned later. One used in the motor industry (often in manufacturer/dealer disputes) is known as the *executive tribunal*, whereby the senior person from each party sits with the mediator whilst the parties present their case; the executives then retire to negotiate a settlement, with the mediator's assistance. Whilst not a true mediation, the principle of having an independent person facilitate negotiations is common to both.

This principle has been developed in non-contentious negotiations, called *assisted deal-making*, or *deal mediation*, which are dealt with in a later chapter. Again, the negotiators use an independent third party to facilitate their negotiations in order to get the best deal (for all sides).

2.1.4 Arb/Med and Adj/Med

This is a mediated outcome of an adjudication or arbitration. The arbitrator or adjudicator does make a binding decision but, with the consent of the parties, does not immediately (or ever) reveal it to them but invites them to negotiate a settlement rather than have the decision imposed upon them. If they do not want to, or fail to reach agreement, the neutral's decision is delivered and is binding. If the parties agree to a mediation, they negotiate, with the help of the now mediator, their own settlement. It is then put in writing, signed and becomes binding. The arbitrator's/adjudicator's decision is not revealed.

2.1.5 Court settlement procedure

In 2006 the Technology and Construction Court introduced a scheme whereby the case judge, after case management conferences and a significant amount of preparation and reading-in, offers to 'mediate' the case for the parties. If the parties agree, but no settlement is reached, the

---

[1] *Hurst* v. *Leeming*, Lightman J 2002.

judge is no longer permitted to try the case and has to withdraw. If no settlement is reached by the parties, the judge offers an opinion on the merits of the case as s/he sees them. Of course the CSP does mean that another judge has to pick up the case and so would duplicate the work of the original judge. However, despite the low take-up, the scheme is considered by the TCC to be a success as they feel that it offers mediation to parties who may not otherwise consider it, and it enables the judge, who has considerable knowledge of the case and the relevant law, to help the parties achieve a settlement without there being an outright winner or loser.

2.2 Resolving disputes through recommendation

2.2.1 Neutral fact-finding

NFF involves a jointly appointed neutral person, who may well be an expert in the field, investigating particular issues and reporting jointly to the parties. This is intended to clarify particular issues that may be causing a blockage owing to ignorance or differences of understanding. The result is non-binding and may still lead to another ADR process for resolution of the dispute.

2.2.2 Dispute review boards

Dispute review boards (DRB) are becoming increasingly popular both in the UK and in Europe. The DRB is made up of three experienced construction professionals, selected at the beginning of the project, who hear disputes when they arise and make non-binding recommendations for resolution. The three members comprise one person selected by each party (and agreed by the other party) and a third selected by the other two who then acts as Chair of the DRB. The DRB only has the authority granted to it by the parties. The DRB's recommendation may be accepted temporarily by the parties until the matter is finally determined by the courts or arbitrator.

2.2.3 Early neutral evaluation (also judicial appraisal)

ENE involves the evaluation by a jointly appointed neutral person, often a judge, of the presented facts from all parties, producing a non-binding opinion on the likely outcome if the case goes to court. It differs from a party obtaining counsel's opinion in that the neutral is jointly instructed and receives submissions from all sides. This is not necessarily a quick or cheap process as it usually involves a significant amount of paperwork, which has to be prepared and then read and considered. The ENE neutral is therefore usually a retired judge! Following the evaluation, the parties attempt a negotiated settlement.

2.2.4   Mediator recommendation

Similarly, a mediator may be asked to make a non-binding recommendation if the parties do not achieve a settlement in the mediation. The mediator can only do so based upon the information given during the mediation and, as some of that may be confidential, the mediator must be very careful to qualify any recommendation and ensure that no confidentiality is breached. Most mediators will resist a request to give any recommendation as it is likely to change the mediator's relationship with the parties and remove the possibility of keeping the negotiations going in the period following the mediation. After all, one of the key advantages of mediation is that the parties construct their own solution.

2.2.5   Conciliation

Although mentioned in the previous section as a consensual process, conciliation under the ACAS rules and in civil engineering disputes also involves a solution being recommended (but not imposed) by the neutral. This is usually preceded by a consensual process where the conciliator tries to get the parties to reach an agreement, but if that does not happen, s/he makes a recommendation for settlement. Again, this is often accepted by the parties until the matter is finally resolved by the courts or arbitrator. The ICE (Institution of Civil Engineers)[2] Conciliation Procedure provides that any recommendation or proposal shall only be binding on both parties if incorporated into a written agreement signed by both parties.

---

[2] www.ice.org.uk.

2.3   Imposed solutions to disputes

2.3.1   Med/Arb and Med/Adj

These are mediations that have a contractual provision for an adjudi-
cated decision if the dispute does not settle. Sometimes the mediator becomes the adjudicator/arbitrator; sometimes it is a separate person. The
Construction Conciliation Group (CCG) in the UK has a time-limited
fixed-price scheme which guarantees a solution by the conciliator making a binding decision at the end of the fixed time, if the parties still
have not reached a mediated agreement. The problem with this process
is that, in the earlier capacity of a mediator, the adjudicator/arbitrator
is party to confidential information that would not normally be shared
if the neutral is to impose a decision. It is very hard for that neutral to
ignore the confidential information provided in the mediation, and so
there is a danger that the parties may hold such information back in
the mediation phase. That would prejudice the success of the mediation
and probably cause any mediated solution to be less efficient.

2.3.2   Adjudication

Adjudication is a form of time-limited fast track arbitration. Although
generally available in any sector, it is a statutory right, and therefore
particularly common, in the construction industry. The adjudicator is
usually a specialist in the area of dispute and will give an opinion,
usually within 28 days, following a papers-only submission or paper
and oral submissions. This is a decision based upon the relevant law
and is binding unless appealed, leading to a final determination of the
disputes by the court or arbitrator.

2.3.3   Ombudsman

There are several ombudsman schemes in the UK, most notably in the
housing and financial services sectors. The ombudsman is usually seen
as the last resort, the parties having exhausted other reasonable means
of resolution. Generally the scheme is funded through a levy on a particular industry and the ombudsman's decision is binding on the industry
member, but not on the complainant. It does not therefore guarantee a
resolution of the dispute!

2.3.4   Expert determination

A jointly appointed expert investigates and makes a binding decision on the issues in dispute. Sometimes this will involve the interpretation of law but most often it will be on technical or financial issues. There are not normally oral presentations, and the expert will normally investigate beyond the parties' submissions before making a decision.

2.3.5   Tribunals

Some sectors, particularly sports and employment, have a tradition of resolving disputes by tribunal. Usually three people, including a chairperson, hear submissions and make a ruling. Lawyers are often involved and the hearings may be protracted. The decisions are binding, with limited grounds for appeal. Because these disputes almost always involve continuing relationships and because delay can cause significant harm (particularly to an individual against whom the complaint may be made), mediation is becoming recognised as a quicker and more effective method of resolution.

2.3.6   Arbitration

This is one of the two 'traditional' forms of dispute resolution to which the others listed are 'alternatives'. Usually one (but it may be three, particularly in international disputes) person with industry-specific expertise, and often a lawyer, will hold a formal hearing and reach a binding decision.

Although there have been attempts to make arbitration a quicker and more efficient process, it is a legally bound process and involves much of the procedure and argument that occurs in a court trial. It is therefore expensive, protracted and inflexible. The advantage over a court trial is that the process is private, and the one handing down the decision is an expert rather than a lawyer and so is likely to understand the issues and the practice more readily than a judge. There is, however, a cost attached to that!

2.3.7   Litigation

Litigation is the traditional form of dispute resolution based, in the UK, on law and precedent. A judge sits and listens to argument on the interpretation of the relevant law as applied to the particular dispute and then makes a decision as to who wins and who loses. A trial can take several days (or weeks) and it is not unusual for the cost to each party to exceed the amount in dispute. Although court lists have shortened (partly because of early resolution through effective ADR), it is still usual for a case to be heard a year or more after proceedings have started. One of the disadvantages of taking a case through the courts is that the result can only be an outright winner and an outright loser, and the redress is financial damages (or an injunction, if that is the issue) and award of costs.

2.4 Why traditional methods fail the parties

2.4.1 Inherent injustice

To my view (as a non-lawyer!), there is an inherent injustice in any process which imposes a decision on the parties. The outcome is usually that there is an outright winner and an outright loser, and the loser generally pays the costs incurred by the winner. But life is not about being totally right or totally wrong. There are endless 'truths' between the extremes. Everyone sees the same facts and events differently, for a whole lot of reasons. It does not necessarily mean that one person's view is more right or wrong than that of the others – just different. So outright winners and losers mean that differences cannot be considered. The person making the decision must believe one person's version of the truth above the others and then make them the outright winner. Not by 10% or 25%, but winner by 100%. Not fair!

2.4.2 Cost and time

Direct negotiation aside, there is no dispute resolution process that is more cost-effective than mediation. Mediations rarely take more than a day and the mediator fee, including preparation, is usually no more than a solicitor partner's day fee but, of course, it is shared equally between the parties. Admittedly each party has to prepare for the mediation and needs to bring a (small) team on the day, but there is nothing in the preparation that would not otherwise be done for another process, so even if the mediation does not settle, it is never a wasted expense. The objection to mediation – that it just adds another layer of expense – is totally misguided. However, one thing is certain: litigation and arbitration (and, increasingly, adjudication) are hugely expensive processes that still, despite CPR, take a long time to achieve an outcome. And one of the biggest expenses is rarely quantified – the amount of management time involved in supporting or defending a claim. The CIF (Construction Industry Federation of Ireland) survey mentioned in Chapter 1 starkly highlighted that fact – 2% of turnover spent in managing disputes. Even if that figure is less in the UK, it still represents a huge sum of money that could otherwise be profit or time spent in wealth-generation.

2.4.3   Adjudication is not the 'saviour'

Adjudication is now well established in the UK construction industry and it has had several significant benefits. It has undoubtedly saved many small firms from extinction and has swept away many disputes that would otherwise have rumbled on for several years. It has given (temporary) finality to disputes and so allowed those involved to concentrate on the successful completion of the project rather than waste resources on preparing or defending claims. It has to be said, though, that the greater the sums involved, the more temporary is the finality, and in high-value claims the outcome of the adjudication is rarely final. Inevitably, it is becoming an expensive process. Inevitably, because it is a legal process and therefore one that has been developing legally since its adoption. Solicitors, even counsel, and experts have become common participants in adjudications and the 28-day limit has become extended more and more. And over all it is still the fact that the adjudicator has to decide who wins, rather than the parties deciding their own outcome.

Having said that, some adjudicators who have trained as mediators do offer a mediated solution, usually after making their decision (but before issue). The dispute resolution spectrum (Figure 2.1) includes Adj/Med (and Arb/Med), whereby the adjudicator becomes a mediator and so enables the parties to reach their own settlement. The adjudication award is always there if the parties are unable, or refuse, to settle in the mediation phase.

One of the good things for the emerging 'profession' of adjudication is the forming of the Adjudication Society[3] and the standard-setting and qualification by other professional bodies such as the Royal Institution of Chartered Surveyors[4] and the Chartered Institute of Arbitrators.[5]

---

[3] www.adjudication.org.
[4] www.rics.org.
[5] www.arbitrators.org.

2.5   The better options for dispute resolution

2.5.1   Consensual processes

This book is about the mediation of construction disputes so it is no surprise that mediation is offered as the best method of resolving construction disputes. But mediation at its best is about cooperation, about working together to achieve a solution to which everyone can say 'Yes', and that may mean a complete culture change. Working from a position of trust and openness may not be the most instinctive way for most construction parties. But it is the way to achieving the best solutions and to fostering good relationships. And good relationships mean good business!

2.5.2   Partnering

There is a section in a later chapter on partnering, but it should also be mentioned here. Partnering is about 'win–win' attitudes: people working together for mutual gain. It is about trust and openness, the mutual pursuit of excellence and the cooperative approach to problem-solving. It is also about having an agreed and progressive dispute resolution process which preserves working relationships and avoids the blame culture. When it works it is great, and there are examples cited later which are stunning successes. Unfortunately, it doesn't always work, often because the adversarial culture is so ingrained and because, when problems occur, parties cannot hold their nerve and resist the temptation to protect themselves (and, in so doing, point their finger at others).

2.5.3   Dispute avoidance

It is not a big step from partnering to creating a culture of conflict management that avoids disputes altogether. Later chapters give some 'rules' and include deal mediation and project mediation, both of which move the mediation skills, and sometimes the process, 'upstream' to the initial negotiations and to the currency of the project. Getting the relationships right means that people work together to resolve problems when they occur. Getting the conflict management processes right means that everyone is confident about what to do when problems arise. It is common sense and it therefore sounds easy, but it actually takes a long time to change a culture. And it is the historic culture of the construction industry that needs changing.

Chapter 2 in a nutshell

- The best resolution is one achieved by the parties themselves. This means negotiation or mediation.
- Doing it the way 'we've always done it' is not a reason to do it again.
- Adopting dispute avoidance programmes is best business practice.
- Partnering works, but it requires a culture change . . . and nerves!
- Progressive dispute resolution processes are the most efficient and cost-effective (negotiation–mediation–DRB–adjudication).
- Just because the contract does not specify mediation (or another dispute resolution process), it doesn't mean parties cannot use it. Parties can agree to resolve disputes in whatever way they wish.

CHAPTER THREE
The Case for the Mediation of Construction Disputes

This chapter argues the case for mediation for an industry that not only has a tradition of arbitration and litigation but has also in recent years readily embraced adjudication. The last chapter argued how inefficient and unjust these dispute resolution processes are, and what follows are the compelling reasons for the construction industry to embrace mediation as the preferred method of resolving disputes in the future.

3.1 Better deals

The fact is that, whether negotiating settlements to disputes or just negotiating commercial deals, these deals are usually better when an independent third party is involved.

Why is it that this dynamic gets better deals through mediation than through other forms of dispute resolution? Consider these reasons:

- No gains are left unknown or unused (see 'Mediator added value' below). Parties who are negotiating will never be totally open and transparent with the other parties. The perception is that, by being totally open, they will be disadvantaged and so the deal will not be as good as one negotiated with their needs concealed. However, a trusted independent third party can be given sensitive information in confidence by all sides, and so will be in the unique situation of seeing where needs are common and where some things are cheap concessions to one yet valuable gains to the other, and will be able to ensure that when the deal is done, everything has been included.
- Mediation provides the opportunity for the whole story to be told, and therefore for wiser deals to be made in the full light of that whole story. Mediation creates a unique forum: a group of people that is gathered with the express intention of finding a resolution to the dispute. It is very unlikely that the same people will have gathered together before – even those on one side, let alone from all sides to the

dispute. It is certain that one side will not have heard the other side's story, or told their own story, face to face, with all the conviction, and even emotion, of people who feel strongly about their version of the truth. Until then they will only have heard their own version and perhaps anticipated the other's.

- Deals are crafted along the whole spectrum of possibilities from one extreme to the other, and from common sense, not legal argument. Both/all parties can be winners. As mentioned in earlier chapters, life is not about being totally right, or totally wrong. We all know that there are many shades between the two extremes. People see the same events and facts differently, for a whole load of reasons (education, culture, age, politics, etc.) and it does not have to mean that they are any more right or wrong – just different. Unlike most other forms of dispute resolution, mediation can take that into account.

- The power is with the parties. All the parties. In mediation everyone has to say 'Yes' to the deal. That means everyone has to take account of the needs of all the parties, not just their own. This usually moves the mediation into an atmosphere of cooperation and away from confrontation. A cooperative solution leaves parties far more satisfied than they would be with any imposed solution. And for lawyers, a satisfied client is good for future business.

- Settlements stick (mediated settlements almost never break down) and parties are therefore often happy to work together in the future. There are few industries in which parties have no further dealings after settling a dispute, and the construction industry is no exception. Quite apart from the fact that the world is shrinking and that paths cross in the most unexpected ways, there is an underlying 'family' bond that runs through most of the industry. Repeat work is common. Good workmanship and trust lead to repeat business even though difficulties may have arisen in the past. Few people can ignore the likelihood of having to work with opponents again in the future. One of the benefits of mediation is that it can strengthen, even restore, relationships. The very fact of recognising that this is a common problem, and therefore parties need to work towards a common solution, creates a different dynamic to adversarial routes. Cooperation usually means understanding, if not accepting, the other's position rather than rejecting it out of hand.

- Even if there is no continued relationship, the satisfaction of a 'fair' deal will prevail. 'Fairness' and 'justice' are words not normally used in mediation. The mediator cannot know what is 'fair' or 'just'; only the parties can measure that. Mediation is often described as creating 'win–win' results, but the reality is that most deals achieved in

30

mediation are tough and take parties into areas where they did not expect to be. If the deals were easy to achieve, the parties would not need a mediator. 'Pain–pain' often seems to be a more appropriate comment than 'win–win'. But even in the most difficult settlements, the fact that parties are seen to share the pain is often enough for deals to be struck that would otherwise not seem possible.

- And those deals are not always only about money. Anything can be put into the settlement pot: staged payments, discounts on future work, guaranteed annual turnover, mergers/takeovers, joint ventures, even each party bearing their own legal costs, and 'drop hand' deals (parties walk away from the dispute agreeing not to pursue it further). Anything (legal) is possible. So the settlement 'pie' is potentially larger and the ingredients greater, and so the parties' share can be much bigger than a money-only deal. Arbitration and litigation cannot do that. They dictate that one party is an outright winner, the other an outright loser, and usually that money must be paid in compensation.

Case study

Boundary dispute where one party (A) maintained that the adjoining party (B) had encroached 5 metres into A's land. A surveyor demonstrated that this was so although it was also accepted that the encroachment was very old (in excess of ten years). Feelings were strong and both parties involved lawyers but they realised that their relationship was being severely damaged by the dispute and chose to go to mediation. A totally new boundary was agreed, some money changed hands and a bottle of champagne was opened to celebrate a creative and cooperative solution.

Litigation/arbitration/adjudication could never have achieved this outcome. One party would have been decreed to be right and the other wrong, and relationships would have been damaged forever. Certainly there would have been no champagne.

Case study

Contractor in dispute with industrial property owner over suspected defective concrete to the external access roads and lorry parks. Some core samples were found to be a weaker mix than others but they were intermittent. The owner required either total replacement of paving with correct concrete mix or a significant financial discount from the contractor's final account. This involved many millions of pounds and, although strictly in the wrong under the contract, the contractor was convinced that there was no problem, so he fought the claim. Each party spent over £100,000 in legal costs before taking the case to mediation. A settlement was reached in the mediation whereby the contractor agreed an extended Defects Liability Period and set up a free site maintenance facility to deal with all the owner's requirements for a two-year period, at a cost to be negotiated when the two-year period had expired.

Litigation/arbitration/adjudication could never have reached this outcome. The likelihood was that the contractor would have had to remove all concrete that did not meet the contract specification, or pay several million pounds, and face bankruptcy as a result. Both parties would have lost.

3.2 Speed and economy

One of the most compelling reasons for taking a dispute to mediation is that it is over quickly and therefore is an economical process. Most mediations are over in a day. For that day an experienced mediator will charge no more than a solicitor partner, as either an hourly rate or a lump sum day fee that includes preparation. Bearing in mind that 80–90% of mediations settle on the day, this is a small cost to achieve a resolution, particularly when compared with the alternatives. Indeed, one of the motivations to getting a settlement at mediation is the high and often uncertain cost of going to trial or arbitration – the indirect cost of management time as well as the direct cost of the legal team and experts.

With the courts becoming more and more positive in encouraging, if not nudging, parties to mediation, the issue is no longer 'Shall we try mediation?' but 'When shall we go to mediation?' It becomes a matter of timing. That will be discussed in more detail later, but it is a serious matter because the earlier mediation is used, the lower the costs involved but the higher the risk (because the information available is less than after disclosure and witness statements). On the other hand, the later mediation is used, the higher the cost but the lower the risk. When to 'press the button' is a serious judgement call for the party's lawyer (see Chapter 4).

3.3 Flexibility in process and outcome

Chapter 2 compared the resolution options available to disputing parties, from 'informal' conventional negotiation to 'formal' litigation. One
of the benefits of mediation is flexibility, both of the process and, as
mentioned above, of the outcomes.

The conventional mediation process is covered in detail in the next
chapter, but the overall principle is that it can be adapted to suit the
needs of the parties and the dispute. Most mediations have an initial
joint meeting, but that does not have to be the case – sometimes parties
are so wound up that they cannot bear to be in the same room. That usually changes as the mediation progresses and it is very likely that joint
meetings are held later in the process. A conventional mediation will
follow the initial joint meeting with a series of private meetings where
the mediator explores what is important to the parties and what their
real needs are. Rarely, the initial joint meeting continues through the day
and the need for private meetings does not arise. Sometimes all that is
necessary is to get the parties talking to each other again, and they are
then able to work out a solution together.

It may be sensible to hold pre-mediation meetings to plan the day
and to get initial issues cleared away. It may also be sensible to hold pre-
liminary meetings – called 'history days' – in more complex disputes,
so that everyone involved can have their say and thus enable the mediation proper to engage a much smaller number of people who can
concentrate on the settlement negotiations, and the future. Mediation
days can be phased to allow further investigation or actions or to allow
specific people (such as experts) to attend for part of the process.

Experienced mediators should be prepared to treat the mediation
process as something that can be tailored to suit the parties and their
particular dispute. Although the normal accepted process for commercial mediation is something that is tried and tested and proven to
work, there can be many variations that make the parties more comfortable with the process and give them the best chance of negotiating a deal. Provided the mediator is always even-handed, anything is
possible.

3.4 Finality of outcome

What mediation offers is finality. All the risk of what a third party might
decide is removed. The outcome is entirely in the hands of the parties,

and if they reach agreement the deal is done. They sign a brief settlement agreement and the matter is over. No matter how confident a party (and/or their legal advisors) may be, the outcome of litigation/arbitration/adjudication is never certain. All lawyers have their stories of 'dead cert' cases going against them because the judge got out of bed the wrong way ... or got out of the wrong bed the right way! Arbitration and adjudication are similar. So much depends on the person who will make the decision, how they are feeling, how they interpret the evidence and how they relate to the parties. So much also depends upon how individuals will present and how they will stand up to being questioned. There are so many imponderables, even with a 'dead cert' case.

3.5   Mediator 'added value'

There are many benefits to having a mediator (a third party who is independent and has no vested interest in the outcome) involved in the negotiation of a settlement to a dispute:

- As mentioned above, the mediator gets to know the whole story, even the sensitive bits that will not be shared with the other party. In this way, the mediator has a unique picture that no one else can see, and so the mediator is able to help parties craft a solution that not only meets their needs but also ensures that no gains are left unused.
- The mediator brings a clear and uncluttered mind to the issues. So often, parties and their advisors wallow in the detail and lose sight of the original problems. The mediator can bring them back to the basic issues and add clarity to their discussions. With most mediations lasting only a day, it becomes necessary for the parties to let go of the detail and seek a global solution. The mediator helps the parties see the 'big picture'.
- In addition to helping the parties focus on the key issues, the mediator will keep asking the question 'Why?', usually to test the statements or offers being made. Reality-testing is a vital tool, and a service to the parties, for it can help parties turn assumptions into facts; it can challenge inflexible positions and ensure that decisions are realistic, and therefore wise.
- An effective mediator is skilled in helping parties use language to best effect by coaching, reframing and avoiding potential loss of face. The mediator's sole intention is to help parties get the best deal, and coaching them to frame offers and responses in positive and

encouraging language can make the difference between rejection and acceptance.

- The mediator helps build an atmosphere of trust and cooperation. The mediator is able to help parties move from entrenched positions into becoming increasingly cooperative as each party's motives are seen to be genuine. No party wants to risk being vulnerable and so possibly being disadvantaged, and the mediator can help the parties build confidence in each other so that moves towards settlement are reciprocated.
- The mediator also helps the parties move from the legalities and rights of their case to a commercial negotiation where deals are made that suit their business. Rights do not generate agreement, only argument. Mediation restores the focus to business needs and sensible solutions.

Case study

This is a great story which will be particularly appreciated by quantity surveyors. A mediator was on holiday in Africa when, riding his camel and enjoying the peace and quiet of the open desert, he came across a group of Arabs in heated discussion. The mediator asked if they needed help and the older Arab explained that their father had left 17 camels to his three sons, to be shared one half to the older, one third to the middle and one ninth to the younger son. With 17 camels, that involved a lot of blood and spare parts. Being creative (and generous), the mediator offered her own camel to make the total 18. So they divided the inheritance up, 9 to the older, 6 to the middle and 2 to the younger son: $9 + 6 + 2 = 17 + 1$ spare. So the mediator was able to take her camel back and ride off into the sunset!

PS I don't think this story is true, but it's good nonetheless!

3.6 Getting off the treadmill

So many parties with disputes soon get to the point where they just want to end the dispute and get on with life. It has taken over their whole existence, and both personal and business lives are affected. It all starts when they have a dispute which they cannot resolve and so consult a solicitor. The solicitor takes the matter over on behalf of the party, and that's it. The party no longer has control of the matter and both time and money disappear with increasing rapidity. By then many parties wish that they had never started the action and just want to put an end to it. So many mediations settle because parties want closure, to put an end to the time and emotion and the grind of being in conflict. No one can foretell the demand disputes make on management/personal time, where people have to defend an action (or their position in taking action) rather than spend their time on creating wealth and enjoying the fruits of their labours.

Mediation provides the opportunity for parties to end disputes with dignity. A good mediator will help parties move from entrenched positions to ones of flexibility where loss of face is avoided and dignity preserved. Deals can be done that parties can live with, knowing that tomorrow brings the opportunity to look to the future rather than still be immersed in the past. For the claiming party the only other way to step off the treadmill is to abandon the case and risk a costs claim from the other side. The defending party cannot even do that. Mediation gives the control back to the parties and allows them to decide the outcome.

3.7 Ongoing relationships

In total contrast to other forms of dispute resolution, mediation offers the opportunity for broken relationships to start being restored and for ongoing relationships to be strengthened. It does not happen in every case; indeed sometimes the parties never want to see each other again. But we live in a small world and paths frequently cross, so achieving an amicable solution to a dispute can also be an investment in the future. Sometimes an ongoing relationship is essential, whether because the area of dispute is in a close community (e.g. specialised technology) or because separation is not an option (e.g. workplace/boardroom dispute). One thing is certain: almost always the disputing parties are pleased (or at least relieved) that the dispute is over and are prepared to talk to each other again. Indeed, the vast majority of my mediations end with the parties having a drink together and some even result in discussions about further business. None of that happens if the case goes to court (or arbitration or adjudication) where the animosity will be fuelled by there being an outright winner/loser (quite apart from the fact that few parties ever want to repeat the experience). Of course, a pleased client from a successful mediation is likely to return to the lawyer that enabled that success, ensuring an ongoing relationship between client and advisor as well.

3.8 *Day in court*

One of the loudest cries against mediation is that parties want/need their day in court. They want to be able to say how it has been for them and they want to vent their anger/frustration/pain. They may even want to humiliate the other party in public. They certainly want to convince the judge that they are right. And what happens? They are asked specific, closed questions that demand little more than a 'Yes' or 'No' answer, and such venting as can be possible usually takes place outside the courtroom. Worse, the other party's counsel will do their best to discredit them and so weaken their case before the judge. Quite apart from the fact that this is invariably an unhappy experience that parties never want to repeat, this is not a 'day in court' in the way that parties seek it to be.

Mediation, however, provides just such an opportunity. Parties can have their say, can show how they feel, can exorcise their demons; and, provided things don't become abusive or deteriorate into bloodshed, the mediator will give them the time – indeed will encourage the parties to make the best of the opportunity. Invariably the parties will approach settlement more comfortably and more quickly if they have had their say at the beginning. It really can be their 'day in court'.

3.9 *Commercial* v. *Legal*

One of the problems with legal disputes is that they are ... legal. The lawyers, naturally, build their best case on legal argument and do their utmost to counter the other party's legal argument. Counsel are briefed in a way that will reinforce each position and, in the end, a judge has to decide which is the more believable.

Unfortunately, this whole process takes the dispute from a commercial issue to a legal one. The problem usually arises from some human failing, or frailty, and it is then enclosed by a straitjacket called the law. It becomes a matter of interpretation, precedent, legal argument and cleverness. Commercial disagreement is transformed into legal argument – necessarily, because that is the lawyers *raison d'être*, but it takes the disagreement into an uncommercial arena. The disagreement is taken from the 'real' commercial world into a world that often seems to have no connection with reality or common sense. The party loses all control of the disagreement; extreme positions are drawn and argued, purportedly in the name of starting from the 'best' (i.e. most extreme) negotiating position. Sometimes this does lead to settlement before a judge/arbitrator/adjudicator has to make a decision, but how efficient or satisfactory that settlement is must be open to question.

And the reason for this is that the basis of argument is flawed. The matter should not be about legal rights and remedies; it should be about listening to the other party's version of the truth and trying to understand why they feel so strongly about it; it should be about uncovering their needs and working out a solution that meets them; it should be about letting them tell their 'story' and creating an opportunity for you to tell yours. It should be about seeing the problem as being a problem for everyone, and the best solution as one in which everyone is able to contribute and say 'Yes'. That cannot be done in court or in arbitration or adjudication. One party wins, another loses. That's the law.

Mediation is an assisted commercial negotiation. The argument is commercial, and commercial people are in control. It gives the parties the power to make the decision over what structure a settlement might have and what the details are to enable all parties to agree. That's why it works.

3.10   *The arguments against mediation*

It is worth ending this chapter with a note on the most common arguments against using mediation to resolve construction disputes.

3.10.1   It adds another layer of cost

It is sometimes said that mediation causes 'yet another layer of costs' and that this is a reason to avoid it. There are at least two arguments against this: by far the majority of mediations settle, so it is a cost worth investing; the major cost is not the mediator but the legal team, and that is a cost that will be incurred in any event if the case proceeds to trial or arbitration. Further, if the case is one of the few that does not settle in mediation, the parties' understanding of each other's case and the drivers behind their positions is usually so much greater that it leads to, at the very least, a simplification of the arguments and, more usually, to a subsequently negotiated settlement. So the chance of the cost of mediation being a wasted cost is so slim that it should not even be considered.

3.10.2   It is too 'touchy-feely'

I suppose 'touchy-feely' is meant to indicate that mediation is not 'manly' enough for real people. The term is the biggest insult that can be aimed at a process that restores humanity and common sense to dispute resolution. It is usually used in a disparaging way, mostly by 'professionals' who do not consider feelings, let alone emotions, to be part of their existence. Fortunately Daniel Goleman[1] renamed it 'emotional intelligence' and so it has become 'respectable'. Mediation *is* about people being allowed to show that they feel strongly about their position, and it is about those feelings being recognised and valued. It *is* about building relationships of trust, both between mediator and party and, in part or in whole, between party and party. It *is* about restoring humanity to a process where otherwise the inhumanity of law would prevail. It is *not* touchy-feely; it is bloody hard work.

3.10.3 Mediation is non-binding so has no teeth

As people have become used to mediation, they have realised that this is a non-argument. As will be seen later, mediation *is* non-binding until an agreement is reached, reduced to writing and signed by the parties. Until that moment the parties are not committed to anything (except to negotiate in good faith) and so it is non-binding. However, once the settlement agreement is signed it becomes as binding as any contract.

3.10.4 Mediation is all about compromise

There are two assumptions behind such a remark: that mediation is about giving in and/or about splitting the difference. Most of the problem lies with the word 'compromise', yet all negotiations involve give and take (i.e. compromise). In some people's eyes compromise appears to indicate weakness, yet we are doing it all the time. It certainly does not mean giving in.

Secondly, mediation is not about splitting the difference between the parties' positions. There is no pattern to most deals that are done in mediation. Sometimes it may be near the mid-point of their claims but invariably the deal is closer to one party's figure than the other's. The deal is whatever the parties say 'Yes' to.

---

[1] *Emotional Intelligence*, Daniel Goleman (Bloomsbury paperbacks).

3.10.5 Mediation is all talk, no commitment

Of course, the fact that most mediations settle is sufficient counter to this statement. Mediation *is* all talk, and the reason it succeeds is that it reopens blocked lines of communication and gets the right people to the table, seeking to understand each other's position and finding a solution that meets their needs. Most disputes occur because of poor communication. Mediation provides a flexible, unthreatening forum in which to put that right. No matter how reluctant parties may be to come to mediation, once there most people grab the opportunity that mediation offers to obtain a settlement on terms they find acceptable. They do sign a mediation agreement that states that parties come to mediation in good faith (i.e. with an honest intent to try to achieve a settlement) – although no one, least of all the mediator, can force a party to mediate in good faith or to reach a settlement. Almost all do.

Chapter 3 in a nutshell

- Introducing a neutral third party to negotiations gets better deals.
- Mediation is quick (usually one day) and cheap.
- Settlements can go far beyond the limited outcomes that are possible in litigation/arbitration/adjudication.
- Mediation settlements are in the control of the parties and not dependent on the whim of a third party.
- Settlement brings finality. Parties can get on with wealth-generation, not wasting valuable management time on pursuing/defending a claim.
- Mediation preserves/restores relationships.
- Mediation restores common sense to the resolution of disputes.

CHAPTER FOUR
Preparing for Mediation

The next four chapters provide a stage-by-stage analysis of the mediation process (preparing, presenting, negotiating, concluding) with the user (party or lawyer) in mind. Most mediators say that few people use mediation to best advantage. This chapter should enable, and encourage, readers to do so. Commercial mediations almost always have lawyers present so there is an assumption (and one of the few rules is that mediators should never make assumptions!) in what follows that lawyers will be present and assume leadership of the team.

4.1   Typical framework

Prior to the day there will be some preparation work and contact between the mediator and the lawyers (and possibly also the parties). On the day, most mediators will meet the parties and spend some time with them in private. At some stage the mediator will bring the parties together in the main room and invite them to summarise the key issues, why they feel strongly about them and how they might be moved forward towards settlement. Some (probably more experienced) mediators will keep that first joint meeting going for quite a long time, perhaps two or three hours; others will break immediately after the opening statements. But at some stage the mediator will hold a series of private meetings with the parties in their rooms. There may also be working group meetings (e.g. lawyer/lawyer or expert/expert) – whatever seems to be useful to help the dispute move towards settlement. Mediation is a flexible process and the mediator should be flexible, and sensitive, enough to vary the framework of meetings to suit the particular case and parties. A typical commercial mediation will have a framework something like that shown in Figure 4.1.

4.2   Stages of mediation

If there is a theory of commercial mediation it is that there are typically five stages (Figure 4.2). When in a mediation, it is probably not so easy

Typical mediation framework

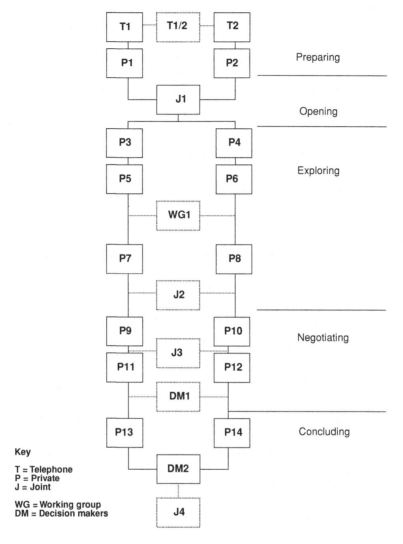

Figure 4.1 Typical mediation framework. Hard lines show typical framework; dotted lines indicate possible variations.

to identify these stages, but looking back from the end it should be easy to see how the mediation flowed through one stage to the next. Sometimes the flow is diverted and the mediation does not run smoothly from preparing, opening and exploring through to negotiating and

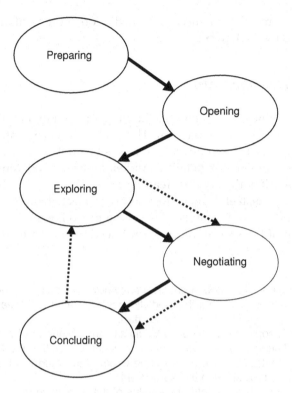

Figure 4.2 Stages of mediation.

concluding. Sometimes it is necessary to return from negotiating or even concluding back to exploring, because a new issue arises or an old one needs further understanding. The process is flexible and can be adapted to suit any case and party. From a party's perspective the five stages become three – preparing, presenting (which includes exploring) and negotiating (which includes concluding).

This and the following chapters take each of the stages and examines its purpose, what the mediator is likely to be doing and how the user can obtain best advantage. By 'best advantage' I do not necessarily mean advantage over the other parties (although they may not be as good at using the process as you are). Mediation moves parties (or tries to) into becoming cooperative, and so gaining advantage over the other is not necessarily helpful. In the end, though, the mediator is there to give the parties their best shot at doing a deal. It is not the mediator's responsibility if the parties foul up (although most do not, despite their seemingly best attempts to do so).

Preparation is the key to success in mediation – preparation by the mediator and by the parties.

4.3 *Preparation by the mediator*

The mediator will want to go into the mediation confident about the issues in dispute and the quantum. This means sufficient information to have a confident grasp of the key issues but not so much that the mediator is swamped with detail. Confidence is a personal thing – some mediators have it with very little reading-in; others, particularly lawyer mediators it seems, feel the need to read every sheet of paper available.

I use a standard confirmatory email along the following lines to give an indication of the sort of information I need to enter a mediation in confidence:

> Greetings all!
>
> Following our exchange of emails, I have now booked [date] for your mediation and assume that the mediation will start at 09.30. Please try to arrive in plenty of time for a prompt start.
>
> I attach a copy of my standard Mediation Agreement that I propose using. I will bring the hard copy with me for signature on the day. Meanwhile, any contact that we have will be deemed to be covered by the confidentiality provisions of the Agreement.
>
> Please let me have a note of the value of the dispute so that we can confirm a fee. I attach my fee guide for your information. It is usual for my fee to be shared equally between the parties and I will email an invoice to each of you for payment seven days before the first mediation date (i.e. by xxx). I presume that the invoice should be made out to you as the parties' legal representatives.
>
> I will be preparing for the mediation on [date] and I would appreciate a summary before then from each of you of 5–10 A4 sheets covering:
>
> > brief history of the dispute
> > key issues in dispute (legal/commercial)
> > details of claims/counterclaims and your respective positions on them
> > matters that are not in dispute
> > settlement discussions and offers
> > suggestions that might help settlement
> > confirmation of who is attending the mediation (and their roles).
>
> You may also wish to provide a 'mediator's eyes only' paper. These can all be emailed to me if you wish, without hard copy backup. Support documents should be brief and, if possible, agreed between you.
>
> I understand that the venue will be at [address] and that [person] will be making the arrangements. Please ensure that the main room will be

large enough to take everyone attending plus my Assistant and me and that each party has a comfortable room as their base. I would also like a flipchart in the main room (and in the others if possible) please.

Finally, I will have an Assistant with me at the mediation. [Her/his] name is [xxx] and [s/he] is attending mainly to gain experience. Of course, [name] attends without cost and will be bound by the confidentiality provisions. I attach [her/his] profile for your information.

Feel free to contact me whenever you wish.

All the best. David

Invariably dates and venue are arranged by the parties' lawyers. The mediator usually gets involved after the parties have agreed a date, although sometimes there will be a call to the mediator to check which dates s/he is available. It will be seen from the email above that most mediators require minimal paperwork. The summary helps everyone focus on what the key issues really are – so often they have been lost in the detail of legal argument and peripheral issues – and the support papers help identify the key documents that enlarge on the summary.

The time taken by the mediator to read-in depends on the case and on the mediator. Three to five hours is normal – probably more on larger or more complex cases. The important thing for everyone to remember (including the mediator) is that substantive knowledge of the issues is a secondary requirement. The primary role of the mediator is to manage the process and give the parties their best shot at doing a deal. A good mediator should be able to successfully mediate a case without any papers or reading-in.

Case study

One construction mediation went ahead without any papers being provided to the mediator, despite a considerable number of requests. When the names of the parties were revealed, the mediator assumed that, because one was an air guidance systems manufacturer, the lack of information was due to the secret nature of the work. The mediation reached settlement on the day and the mediator recognised that, after an hour or so, the lack of knowledge ceased to be a problem. He also realised that the lack of information had nothing to do with secrecy and everything to do with the lawyers being 'too busy'. At one level their position was justified for the case was settled and their faith in the mediator confirmed. At another level their laziness set a dangerous precedent.

Once the papers have been read, the mediator will telephone the lawyers and run through the mediation process (all mediators have their preferences and so it is worth doing this even with the most experienced lawyers), settle any queries and discuss some of the background of the dispute and the personalities involved. This is useful 'colouring' of the canvas in preparation for the mediation and it is likely that the lawyers will be more open about the case and their clients in this sort of private conversation than would necessarily be the case on the day.

4.4 Preparation by the parties

4.4.1 When to mediate

In some cases even agreeing to mediate can take time. There is, of course, a 'right' time to mediate, being the time when parties and their advisors feel they know sufficient for the risks to be minimal yet have avoided excessive legal costs and management time. Mediating early, even before legal proceedings are commenced, keeps costs to a minimum, but the risks are greater. Risks that not all the important information is available. Risks that the 'smoking gun' that is the dream of all litigators (but rarely, if ever exists) will be missed. Risks that the deal struck in the mediation will not be the best one available. But on the other hand, there is nothing worse than mediating a case in which the costs exceed the claim (or the settlement). It becomes a mediation about who pays the costs, not the claim. So when to mediate should be a commercial decision. It is the hidden costs of 'wasted' management time that should be the main driver. Time spent in justifying or defending a claim is time lost for wealth-generation.

Some parties will use this as a tactic, either delaying and so putting the other party under pressure of mounting costs, or being premature and so putting the other side at risk of costs sanctions by refusing to mediate at that time. The UK courts are very positive about commercial cases going to mediation so parties will have to consider it sometime. This is a powerful argument to reluctant parties for it is not a matter of 'Shall we mediate?' but much more one of 'When shall we mediate?'

4.4.2 How long should the mediation take?

It is usually the parties (or rather their lawyers) who decide on the length of the mediation. Most take a day but construction disputes are usually complicated, have several parties and involve a lot of detail, so it may be more appropriate to book two days. The trouble is that if two days are allotted, parties will fill them. Serious negotiation will be delayed until the second day (and usually the afternoon) and deals will be done in the evening, no matter how long has been set aside. On balance, it is far better to set aside one day and make sure the evening is free in case more time is needed (and to use co-mediation – see below).

4.4.3   Who to choose as mediator

Recent research[1] of mediation users has established that patience, optimism, clarity of communication, and ability to quickly build relationships and inspire trust are the essential characteristics of an effective mediator. Mediators would add having good listening skills, stamina and a sense of humour! Where do you find such people?

There are three routes to selecting a mediator:

- use a provider
- go direct to an independent
- use a scheme

There are some 50 mediation providers in the UK, most accredited with the Civil Mediation Council (CMC) and some of whom are listed in the appendix. The CMC requires the provider's panel members to have minimum standards of training, professional development, insurance and systems for complaints and the like. There is currently no regulation as such but at least the CMC ensures transparency and a minimum level of professionalism, which many of the established providers exceed. The advantage of using a provider is that they usually offer an independent choice of mediator (and so avoid potential argument between the lawyers) and some support to less experienced users. The disadvantage is that they add to the cost, although in most cases not significantly, and promote their favourites. Nowadays, with their increase in experience of mediation, many lawyers see providers as no longer adding value; having compiled their own list of mediators over the years, they are able to draw on other lawyers' experience when proposing a mediator.

Choosing an independent mediator may well be the cheapest route, although in the scale of things, the mediator fee is unlikely to be significant

---

[1] Amanda Bucklow's research, *The Skills, Attributes, Behaviours and Strengths of Excellent and Effective Mediators.*

compared to the cost of each party's legal team. Except perhaps for the lowest value cases, mediators should never be chosen on cost alone but on experience, reputation, specialism perhaps. The trouble with using an independent is that they are usually nominated because someone had a good experience (or even an experience that was not bad) and so are considered to be safe, whereas they may not be the best mediator for the particular case. At least some of the providers, especially ones who also train mediators, are able to refresh their panels and spot and develop the new stars.

There are now many mediation schemes in existence, particularly for low-value cases. In addition to the mediation schemes offered by most courts, the Construction Conciliation Group (CCG) offers a fixed-fee, fixed-time package with a binding decision if the case does not settle in mediation and the Royal Institution of Chartered Surveyors runs a scheme for the settlement of neighbour (boundary) disputes which is a three-stage fixed-fee process offering report, mediation and expert determination. These schemes serve a useful public service and offer parties a way out of their disputes without running up huge fees, but they are for the small-value disputes and so not relevant to mainstream mediation.

Sometimes, in the very large high-profile construction mediations, a 'beauty parade' of mediators is staged, where potential mediators are interviewed. Sometimes parties want a Big Name as mediator (someone who has not necessarily been trained and accredited but whose gravitas is seen as being important). Often this is a form of game-playing and/or ego-massaging and just clouds the issue. Similarly the trend of choosing a specialist mediator – it might give comfort to the parties that the mediator comes from their environment and so will understand their problems and speak their language, but there the benefit ends and the possible disadvantage of the mediator getting seduced by the detail because it is the mediator's field of expertise is very significant. There is only one thing worth remembering when choosing a mediator: a good mediator can mediate anything. So find a good mediator.

One further comment on the choice of the mediator – beware of serial users. Lawyers, insurers and some other repeat users of mediation have their favourites: tried, tested and with good settlement rates. The danger is that the mediator can get comfortable, used to the party's style of negotiation and their tactics. Worse, the party can get used to the mediator's own style and tactics, and so manipulate her or him to their own advantage.

Case study

A serial user insurance company appointed the same mediator on five alleged fraud cases over a period of two months. Although some members of the claimant insurer's team changed, the decision-maker and lawyer were common to all cases. By the fourth case the mediator knew them well but unfortunately he was seen by the defendant to appear more relaxed and friendly with the claimant party and this was (wrongly) interpreted as being partial. Although they had been aware before the mediation of the mediator having several cases with the claimant (and this might have caused them to be extra sensitive in the matter), the perception of the mediator appearing not to be even-handed became an issue. Appearance of even-handedness on the part of the mediator is crucial. The mediation continued to a successful conclusion but it was a lesson for the mediator not to allow familiarity to challenge the key requirements of neutrality and impartiality.

Ironically in this same case the mediator had become familiar with the claimant's negotiation strategy of playing 'hard ball' for most of the mediation and then achieving a quite reasonable solution at the end. It was very tempting to reassure a despondent party that all would come right in the end, without undermining the claimant's tactics. Familiarity is a problem best avoided.

4.4.4 Mediator fees

There are a variety of ways in which mediators charge for their services:

- lump sum, all-in fee – this may vary according to the value of the dispute and with the number of parties involved
- day fee plus preparation at an hourly rate
- hourly rate

My preference is the first because everyone knows the cost before the mediation takes place but the challenge then is to contain the amount of time spent on preparation. Most all-in fees will contain a set amount of reading-in time (probably three to five hours) and, if the mediator is one who needs to read every piece of paper provided, it can be a challenge for the mediator to contain the preparation within the allotted hours. Of course, if such a mediator is chosen on an hourly rate basis then time spent in preparation is not an issue (at least not for the mediator) but the party has little or no control over the hours spent.

Case study

An American mediator successfully mediated a large two-party case where the parties acknowledged that they saved millions of dollars by settling in mediation. Following settlement, the mediator wrote to the parties requesting an additional fee that reflected the savings made. One paid, one complained and went to the press. The mediator's reputation was not enhanced by the outcome!

Actually success-fee mediations have been known in the UK, whereby the mediator (or the mediator provider) only gets a fee if the case settles. This is a practice that most mediators would consider to be unethical because the mediator then has an interest in the outcome and so cannot be considered totally independent.

4.4.5   Co-mediation

It is worth mentioning co-mediation at this stage although it is dealt with in more detail in Chapter 8. Co-mediation is underused in the UK but – I think – it is ideal for multi-party cases (and therefore suited to many construction cases). It involves two experienced mediators working together as a team. This not only brings another mind to the problems but also enables the time to be used more efficiently as the mediators can work separately in private sessions. One of the weaknesses in mediation is the 'idle time', particularly in multi-party mediations. If a sole mediator spends half an hour with each party in a six-party mediation, it is three hours before the mediator is able to return to the first party. This does nothing for momentum or energy or indeed for keeping the parties engaged in the process. Co-mediation can overcome that by the mediators separating and seeing parties individually. It results in a much more efficient process at very little extra cost.

There are a few 'rules' to consider when choosing co-mediators:

* Use an established pairing. To be most effective, the mediators must be familiar with each other and able to work in harmony. Parties do not want two competing egos mediating their case.
* Decide what is needed from the mediators in style, experience and combination. For example, non-lawyer/lawyer, female/male, specialist/generalist.

- Interview the pairing to be sure they are suited to the case and the parties.

4.4.6 Assistant mediators

Having been chosen, the mediator may quite possibly bring an assistant with them on the day. Indeed they should bring an assistant because it is the only way that newly trained, and therefore inexperienced, mediators can safely get experience before they become lead mediators. From the mediator's perspective it is useful to have another mind and another pair of eyes to observe and comment, particularly in the short breaks as the mediator shuttles between private meetings. From the parties' perspective, having another person with a different background and experience can only be an advantage. My ideal assistant (with apologies to all the male assistants who have accompanied me over the years) is a female lawyer so that between us we cover male/female, business/lawyer dimensions.

Of course, as will be seen from the confirmatory email earlier in this chapter, the assistant comes without cost to the parties. The mediator may pay their expenses but that is a private arrangement. And, of course, the assistant is bound by the confidentiality provisions of the mediation agreement and will, with the mediator and the parties, sign the agreement on the day.

4.4.7 Conflicts of interest

The matter of conflicts of interest is increasingly relevant, particularly to lawyer mediators who belong to large firms. I consider it to be largely nonsense and, in the context of mediation, to have gone to ridiculous lengths. The role of the mediator is to give the parties their best chance of getting a settlement. S/he has no interest in the outcome, and 'favouring' one side would have little or no effect. The mediator does not make a decision – that is the parties' role. The fact that a partner in another office of the mediator's firm did some work for one of the parties several years ago seems to me to be totally irrelevant and should never lead to a mediator being conflicted out. Of course, if the lawyer mediator has acted for one of the parties in the past, that might be relevant and should be disclosed to the other parties. But it is only where the mediator's independence and even-handedness could be perceived to be challenged that conflicting out becomes relevant. Fortunately, for me, as a non-lawyer, the situation rarely arises (although I was once asked if it mattered that I had trained one of the lawyers who would be representing a party in a mediation where I had been appointed – if that were an issue I would be conflicted out of half the mediations that I do!).

4.4.8 Documents

The mediator will usually give a guide to the amount and type of documentation required for the mediation (see the confirmatory email). Most will require a summary – probably no more than five to ten A4 sheets covering

- a brief history of the dispute and the people involved
- the key issues in dispute
- details of the claims/counterclaims and parties' respective positions
- matters that are agreed/not in dispute
- details of any settlement discussions and offers
- suggestions that may help in achieving a settlement

This summary will normally be exchanged simultaneously between parties and the mediator. Lazy lawyers will often send the mediator a copy of the pleadings in lieu of the summary. Pleadings are not an adequate substitute because they deal with the dispute in a legal way, using legal language.

It will be seen from the above that the summary is in part a more general overview of the case but also covers many matters outside the scope of any pleadings. The summary is also a very useful way of getting the lawyers, and parties, to focus on what is really at the root of the problem – something that often gets lost with time as the detail takes over. The tone of the summary document is important. It should recognise that mediation is ultimately a cooperative process and the summary provides the opportunity to start that process. Just because a party feels that their case is strong does not mean that they need to rubbish the other side's case in their summary. A clear, brief summary covering the areas listed above and recognising that this is a joint problem to which there needs to be a joint solution is the type of document that is most helpful to the mediator and most encouraging to the parties.

Case study

In a construction dispute involving the erection and fitting out of agricultural buildings the defendant's mediation summary was prepared by counsel. It contained phrases such as:

'you will lose this point' [several times] 'ill-conceived' 'fatally flawed' 'incompetent'

'inexperience' [the other lawyer] 'we will be paying nothing' [they did!]

throughout the document and was presented on the Friday night before the Monday mediation. The effect on the claimant was that they cancelled the mediation, and the mediator spent much of the weekend persuading them to continue. They did agree to continue and, predictably, counsel arrived late on the morning of the mediation, keeping everyone waiting whilst he briefed his team. His actions were undoubtedly contrived to wind up the claimants, and it took several hours to repair the damage and help the parties to move towards settlement. The mediation did settle, but this was made so much harder by the arrogance and insensitivity of the defendant counsel (who left feeling very pleased with his performance but saying 'I've just lost a £30,000 court fee').

Sometimes parties may wish to provide a brief 'for the mediator's eyes only' paper in addition to the above. This may include comments on relationships and other matters that could be difficult to include in the exchanged summary.

In addition to the above there will be a bundle of supporting documents. Ideally this is a bundle that is agreed between all the parties. Otherwise each party produces their own, and this leads to a lot of duplication. In construction disputes particularly, the extent of documentation can be a real problem. Construction disputes are always full of detail and the supporting documents can be vast: five to ten A4 summary sheets can be accompanied by five to ten lever arch files of supporting papers; or even five to ten boxes of lever arch files of supporting papers! (I do wonder sometimes if the often vast quantity of unnecessary paperwork that I receive has more to do with a billing motive than considered usefulness!) The last thing that a mediator really needs is files of daywork and time sheets, invoices and site records. Summaries, perhaps; individual sheets, definitely not. Mediations never settle as a result of discussions on detail. Those will have/should have happened beforehand – and will have ceased to be useful – hence the mediation. In the end global deals are done in mediation arising from discussions that become increasingly general and 'big picture'. So the rule is to keep the documentation as thin as possible. The last thing parties should need is for the mediator to get immersed in the detail. The mediator's role is to get everyone else out of wallowing in detail and into commercial negotiations. It is not a bad idea to keep asking the questions 'Does the mediator really need this? Is it important? To whom is it important and why?' It can be surprising how few documents fall into the 'vital' category.

4.4.9   Where to mediate

The date and the mediator having been sorted, the next thing is the venue: preferably with easy to access rooms with natural light, outside areas and decent catering. It used to be assumed that a neutral venue, probably halfway between the parties' offices, would be best and fairest. Nowadays the majority of mediations take place in one of the lawyers' offices and the possibility of 'home advantage' (where the host party is likely to benefit) rarely arises. There are several advantages in using lawyers' offices, especially in a city. Public transport and access are usually good and there is always a wine store somewhere on the premises so that settlement of the dispute can be celebrated in the proper way. The main consideration must be that it is the environment best suited to helping the parties reach a settlement. That means good-sized and comfortable rooms for each party that are soundproof and near the main meeting room; a main meeting room that comfortably seats all those attending plus the mediator(s); access to fresh air and space to get some exercise and/or go for a walk. It also means a catering facility that is sensitive to individuals' needs and can produce food at 19.00 as well as at 13.00. And a constant supply of tea/coffee/water/biscuits. That should keep everyone happy and energised no matter how long the mediation takes.

Some mediators (and this probably reflects who trained them!) use flipcharts and ask for them to be provided in each room. I find them to be an essential tool, providing a neutral focus and visual display for issues and quantum.

One other matter that may affect the location of the mediation venue is whether or not a site visit is valuable and, if so, whether or not it should take place on the day. If this is required, the venue needs to be near by. In many cases site visits are very useful for the mediator. Although the mediator is not there to judge, it is very useful for him/her to have a picture of the site/project, hear first hand the issues in dispute and be able to place the subsequent discussions in context. If a visit is not on the day then the venue is unaffected.

4.4.10   Who attends?

Most mediators would urge parties to keep their teams 'lean and keen' (few and committed):

- There must be the highest level of decision maker that is possible; often someone who has not been intimately involved in the project and who therefore can stand back from the emotional pressures is a good choice for they are likely to make a more dispassionate commercial settlement.
- The party's legal advisor is required, whether external or in-house.

After that, as few people as possible for the following reasons:

- Experts rarely help move the dispute to settlement at a mediation. They have done their report, possibly met with the other experts and agreed common ground (or not), and it is highly unlikely that they will alter their position in a mediation. Usually they do nothing and say little. But there is a danger of their becoming a blockage to settlement because they believe their version of the truth is correct – just as the other side's expert believes their own version of the truth. So best to leave them behind. They can always be on standby for a conference call if the need arises. Most mediators, when faced with experts in a mediation, will take the opportunity to put them in a room together and, whilst not exactly locking the door, will try to keep them out of the way for as long as possible.
- More and more mediations have counsel present. This is another level of expense that is rarely justified. Few counsel can resist assuming leadership and doing all the talking. The worst will gag their clients and treat the mediation arena as an extension of the courtroom. Generally, the counsel will have given their opinions, and these opinions, like experts' reports, will invariably disagree and no amount of discussion in the mediation will change that. Mediations are settled not on legal argument but on commercial imperatives. Counsel are best left for after the mediation, and then only if it does not settle (which of course most do). Having said all that, there have been some mediations where the presence of counsel has been a real force towards settlement. The trouble is, you only know that after the event!
- Those who were at the 'coal face' (site agents, foremen, etc.) can also be a blockage. It may be important for them to attend the mediation so they can tell their story, but there is always a danger that they will get annoyed with other people's versions of the truth and become

emotionally attached to what they see as justice (and therefore be a force against a commercial settlement). One option may be to have one day for those that lived the project on a day-to-day basis (a 'history' day) and then have another day for the decision makers and their legal team to negotiate the deal.

- Consultants (architects/engineers/surveyors/project managers) may have their own agendas. Any possibility of blame being attached to them will invariably cause a defensive reaction, the spectre of potential professional negligence claims being ever present. And, frankly, it is unrealistic to expect an architect or engineer to issue a further extension of time or confirm some contested variations just to enable a mediation to settle. The architect or engineer has probably spent months, if not years, protecting the client's interests and to expect that to change in a mediation is unrealistic. So, whilst it might be considered to be necessary for consultants to attend, be realistic about their usefulness (and their own agendas).

Face is a real issue in disputes and it will be appreciated from the above comments that there is a danger of people making statements which they may eventually regret or finding themselves in a position of being challenged and their position questioned. One of the mediator's skills should be to anticipate and avoid such situations or at least ensure that there is enough 'wriggle room' for people to avoid losing face. One way of avoiding these situations is to choose a team for the mediation that will not create such potential blockages. As mentioned in the last chapter, deals with dignity are every mediator's aim.

One danger when deciding on the mediation team is that of tit-for-tat. One party brings their expert; the other feels the need to reciprocate. If one party brings counsel, it is a brave lawyer who advises against the other bringing theirs in return. And so on. There is of course some need to balance the teams so that one party does not bring a team of accountants whilst the other brings a team of technicians, but lean and keen should be the guide.

4.4.11 Authority

The decision maker is the key player in a mediation. S/he decides whether to say 'Yes' or 'No' to the deal that is on offer. S/he comes with authority to settle on behalf of the company/family. The mediator will have checked that the decision maker does have the necessary authority to settle (although most people come with a limit – which is often insufficient and has to be revised) but will also ensure that lines of communication are open if limits become challenged, particularly out of normal working hours. The best situation is for the defendant decision maker to come with authority to settle up to the full value of the claim, and for the claimant decision maker to be prepared to drop hands and walk away if that is the best deal possible. Of course, both are unrealistic situations and so the mediator has to accept that full authority to settle is a very rare thing indeed. Even sole practitioners invariably have a spouse at home who needs to agree to the deal on offer. In fact there has been more than one case where a sole practitioner has come to mediation under instruction to settle the dispute 'or don't come home', because the dispute has so taken over their personal as well as their business life. As mentioned in the last chapter, getting off the treadmill is often the most powerful motivation to settle a dispute in a mediation. So the rule is: the highest level of decision maker with the highest level of flexibility to settle.

Sometimes there are parties who come to mediation with little or no authority. Public bodies, government departments and so on usually have to get a settlement ratified by a committee or officer, and most parties in a mediation with such bodies accept this as part of the deal. It is important that the need for ratification of a settlement is understood from the beginning, and most mediators would link that with an undertaking from whoever is representing the body to positively advocate any deal that is reached in the mediation. Otherwise it is too easy, in these days of blame culture, for such representatives to decide nothing and take no responsibility.

Case study

A construction mediation in Asia had a state-owned company as defendant and an international construction company as the claimant. It was accepted from the start that the government negotiators would take no responsibility for making a decision and so the mediation was structured with breaks (some an hour, some half a day, some several weeks) so that the negotiating team could go back and report to the minister, who in turn went to those at the top of the government structure and, on receiving further instructions, returned to the mediation for the next phase of negotiation. (It was stated at the time that one reason for this was that if the wrong decision was made they might lose their jobs; if it was a bad decision they might lose their liberty; and if it was a really bad decision they might lose their lives. In those circumstances, who would want to make any decision? I am not sure if such stories were true and, even if they were, as this case was many years ago it is very likely that such a culture no longer exists.) Amongst other things, this demonstrates again the flexibility of the mediation process.

4.4.12 Who presents?

Most mediations start with an opening session with everyone present. Most mediators will encourage the parties to use this time to tell their story. Invariably the lawyers will lead, and outline the legal context of the dispute, but the really important statement should be from the party. It is, after all, their money/loss that is the cause of the dispute and they are going to speak with much more feeling than any legal advisor, no matter how eloquent that person might be. Most mediators will invite everyone present at the mediation to speak but, unless they are prepared for it, most parties will stay quiet and let their lawyers do the talking. This usually changes as the mediation progresses, which is a strong reason for keeping the opening session going for quite a long time (at least one or two hours).

The practice should be for the party to speak out from the start. Mediation gives the parties their 'day in court' (see previous chapter); it is they who know what really happened and it is they who feel strongly about their version of the truth. Better therefore that the legal advisor briefs their party to have their say, and to make sure that they make the most of this unique occasion. Everyone else in the team can fill in the gaps if necessary.

4.4.13 Dry run?

Part of the effective preparation for a mediation, particularly if the party is to take the opportunity to have her or his say, may be to have a dry run. Usually the legal advisors and their clients will have a conference on one of the days before the mediation date, so why not do a practice run of the opening statements, critiqued by the rest of the team? It means that the balance between legal and commercial emphasis can be agreed, and that everyone can make the best of the opportunity that the opening session provides to show strength of feelings and to inform the other parties of their alternative versions of the truth. The first open session is a unique gathering – the same people will not have gathered together before to discuss the dispute, and although the lawyers will each have anticipated the other side's approach, they will not have heard it from the people who are experiencing the problem, and neither will they have heard their legal and commercial interpretation of the facts.

The opening session provides a great opportunity to inform and persuade, and it is worth preparing for it well. This is particularly important if the party – or others who might be in the witness box if the dispute goes to trial – is nervous or unused to speaking in public. Often lawyers see mediation as an opportunity to check how the 'opposition' perform, and their willingness to settle in mediation, or to risk court, will be influenced by what they see and experience in the mediation. A dry run would enable the team to recognise the reality of this and to coach parties to give their best performance in the mediation.

Case study

Parties in a construction mediation gave great importance to their experts' reports, which (of course) gave totally different interpretations of the situation. Both experts attended the mediation and it became very obvious that one of the experts would not perform well if examined in the witness box. For some reason the lawyer had not met the expert before the day of the mediation and so this had not been appreciated before the mediation – a situation that was undoubtedly due to poor preparation. Settling in the mediation became an imperative for this party, and they did. (Ironically, this goes against what I said earlier about not bringing experts to a mediation, although the mediation was likely to have settled anyway – because most do – and it might well have been on the same terms.)

4.4.14 Pre-mediation meeting

In England it is unusual for parties to request a pre-mediation meeting. In Scotland, however, it has become the custom and the advocates for this early meeting (which is generally only one or two hours long) say that it pays real dividends on the day of the mediation:

- Everyone 'hits the ground running' so the time on the mediation day is used more efficiently.
- There is a belief that the settlement rates are much higher (probably 90%) as a result.
- It ensures that all the essential information has been exchanged before the day of mediation.
- Where action has not started, it usually results in an agreed statement of facts.
- The mediator is able to influence (and usually reduce) the amount of support paperwork that accompanies the statements and the size of the teams attending.
- It is often the first time that the lawyers have had face-to-face discussions, and that not only reinforces a relationship (which is usually a good thing) but can also lead to some degree of cooperation and even agreement.

However, the pre-mediation meeting does add another cost, and as most mediations settle on the day it is probably difficult to justify in a culture where such meetings are rare.

Having said that, I have already quoted a case where pre-mediation meetings were a real asset.

Case study

It was only a two-party dispute but the gap was around £80 million and it was agreed that the mediation day should have only the key decision makers and their legal advisors (a total of four on each side). The pre-mediation history days were spent allowing the 'coal-face' team members to state their case, each of the two pre-mediation days being divided according to the issues being discussed and each party having their allotted time to make their case. The two principal decision makers sat with me (the mediator) and heard the presentations from each side. By the time of the mediation day, the decision makers on both sides were fully aware of the issues and of the different versions of the truth, and the day itself became centred on a purely commercial negotiation.

The main purpose of a pre-mediation meeting is to get the procedure agreed and to sort out any queries so that the mediation day itself is efficient and the necessary information is available to all.

> ### *Case study*
>
> A mediation involved an overseas party, who flew in to the UK the day before the mediation and tabled, to everyone's surprise (including his own lawyer's), a further claim (in addition to replacement of faulty goods and loss of profit) for restitution of property after the removal of faulty equipment. The party had no supporting documents because the work had not (yet) been carried out. Needless to say, it was not treated very seriously by the defendant party and the mediation did not settle because they would not take account of the new claim. Indeed no offer of settlement was even tabled because the respective figures were so far apart. Subsequently the restitution work was actually done and the costs incurred were reimbursed as part of a deal reached in another day of mediation (the defendant insisted on reducing the settlement figure by the cost of the first abortive mediation). The lesson to everyone was that, whilst it is accepted that it is more difficult with parties who are based outside the UK, there had been insufficient time allowed by the claimant's team for proper preparation.

One of the advantages of mediation is that it is a flexible process and so it can be adapted to any situation. Although most mediations follow the framework noted earlier in this chapter, designing a process that suits a particular circumstance or party is always a possibility. In such circumstances, a pre-mediation meeting with the lawyers, and possibly the parties as well, would be valuable. The rule, in a process that has no (well, few) rules, is that anything is possible (so long as the parties agree).

4.4.15 Pre-mediation contact

From the moment of appointment there will be some contact between the mediator and the lawyers. It is a time when the mediation agreement should be circulated so that the lawyers can agree the terms before the day of mediation and so that the conditions, particularly the confidentiality provision, can operate from the very first contact. Nearer to the mediation date and after reading-in, it is normal (or should be) for the mediator to make more formal contact with the lawyers, and sometimes with the parties (see Figure 4.1). The appendix contains a typical checklist that the mediator might use. The purpose of the call is to

- raise any questions arising from the reading-in
- outline the process proposed for the mediation day – different mediators have different styles, and it is sensible for the mediator to confirm how the mediation will be run and what is expected of the lawyers and parties
- get information on any special needs, time limitations or other 'domestic' issues
- answer any questions that the lawyers may have
- get some background to the dispute and to the personalities involved
- understand the circumstances of any previous settlement discussions and offers, and why they were unsuccessful
- get a feel of what is important to each party and how it might settle

More than anything else, this is an opportunity for the mediator to start the process of building a relationship of trust with the lawyers, and through them the parties, which is essential to the mediator's being effective on the day.

The lawyers should prepare for this call. It is worth thinking about how it can be used to best effect. What is the message you want to feed to the mediator? What are the questions you need to ask, the comments you want to make, the impression you want to leave? Relationships are two-way, so you are able to use (dare I say manipulate?) this early contact to best advantage.

4.4.16 The mediation agreement

As mentioned earlier, the mediator or mediation provider will normally send a copy of the mediation agreement to the lawyers well in advance of the mediation. A typical agreement is included in the appendix but it has to be said that this is a very brief form of what some providers have turned into a multi-page document (with supporting explanatory notes). I always feel that the simplest document is the best (and this applies to the settlement as well as the mediation agreement). It is important to set down the principles that rule the mediation in the mediation agreement:

- confidentiality
- without prejudice
- negotiate in good faith
- authority to settle

After that, in my view, it becomes unnecessarily complicated.

Occasionally, but rarely, lawyers will want to amend the mediation agreement or even use their own agreement. The mediator should really understand why this is necessary because the standard agreement will have been used successfully many times and a change to/from the standard could be dangerous. Worse, it could be a tactic by one lawyer or party to gain advantage over the other.

Case study

A construction law firm produced their own mediation agreement 'based upon their experience over many mediations'. Essentially the main difference from the standard agreement was the provision that, if requested by one party, the mediator would give a recommendation for settlement if the mediation was unsuccessful. This actually took the pressure off the party to reach a settlement during the mediation and their energy was devoted to influencing the mediator to give a recommendation that was favourable to them. In retrospect, if a clause is included in the mediation agreement requiring (or requesting) the mediator to make a recommendation, it should be at the request of all parties, and then only if the mediator agrees.

Sending the agreement out early should mean that on the day it is just a matter of the mediator getting it signed, which will usually happen before the first joint meeting.

4.4.17 Risk analysis

It is essential that parties come to mediation having done a comprehensive risk analysis, so that they are fully aware of the alternatives to settling in mediation, and the range within which they can confidently negotiate. This analysis will inevitably change during the mediation but it must be done beforehand and so provide a sound foundation for what follows. At the very least the risk analysis should include:

- Lawyer's percentage estimate of the litigation risk: What are the best, worst and most likely outcomes if the dispute goes to court? Of course, there is litigation risk even in the most certain cases. All lawyers have stories about the most sure-fire case that went against them because the judge got out of bed on the wrong side This analysis should enable parties some flexibility in their negotiations even in the strongest of cases.

- SWOT (strengths/weaknesses/opportunities/threats) analysis – a comprehensive analysis of the case before going into the mediation.
- BATNA (best – or better – alternative to a negotiated agreement): What will happen if a deal is not done in the mediation? At what stage does that become preferable to the best that can be achieved in the mediation? Some would term this the 'walk-away point', and it is essential to any negotiation. All negotiators need to know the point at which they say with some assurance 'so far and no further'; if they don't, they are likely to concede more than they should and do deals that are inferior to what is possible.

And possibly:

- Decision tree analysis – this, based upon percentage risks, gives the 'value' of the dispute at a particular time. An example is included in the appendix[2] and many negotiators and mediators consider it to be an invaluable tool. Three notes of caution though:
  - It is entirely theory and is based upon spreading the risk. If the case goes to court it is likely that there will be a 100% winner and a 100% loser, so the decision tree percentages become irrelevant.
  - The danger is that the figure produced by the decision tree becomes fixed and ultimately a blockage to settlement – 'the calculations give an answer, therefore it must be right'.
  - It cannot take account of any emotional factors in getting a settlement. The decision tree gives a value for the case at the time and other factors need to be applied for it to be of value to the negotiator.

Case study

This was not a construction case but a situation witnessed in America. The mediator carried a laptop into the private meetings and did a decision tree analysis with the claimant and with the defendant (an insurance company). Both sides were impressed with the process and it enabled them to move their position significantly. Unfortunately the decision tree figures were different for each side

---

[2] Thanks to John Clark (www.mediation-negotiation.com).

> and, although they had moved significantly closer, they stuck to the answer that the decision tree had given. It not only took a long time to get each party to move from that position; both sides became cross with the mediator because he had got them to move so much yet they still had not done the deal. It did settle, but the bad feeling all round was considerable.

In all the above, two 'rules' should apply:

- Always be prepared to revise the analysis as new factors or circumstances emerge.
- In addition to your own analysis, do all the above from the other party's perspective. How does it look from their position? What can you learn from it that might help your negotiations or give you insights which can be used to advantage?

There will be more written about negotiation in Chapter 6, titled (would you believe?) 'Negotiating at the mediation', but there is one essential principle that should underlie all preparations for any negotiation – that is to understand what the needs and drivers (pressures) of the other side are. If you don't know then ask. When you do know what they need and what is important to them, you can construct a deal that meets those needs. If your knowledge is one-sided then the deal will be difficult to achieve and may not be as comprehensive as it could be. The mediator will often use the term 'needs not wants'. It is not what a party is claiming that is important but the needs that underlie the claims.

4.4.18 Anticipate the settlement

Part of sensible preparation for a mediation is envisioning how a settlement might look. Some mediators, even some lawyers, have a template to use when settlement is reached, but these can only really cover the general clauses. Bearing in mind that people are usually most weary at the conclusion of a mediation, giving thought at this stage to the possible structure of a settlement, the elements from which it will be composed and the details necessary to make it work will save time at the end and ensure that the deal has been properly thought through. Some mediations do go on late, even into the small hours, and people cannot be thinking at their best at such times, yet it is essential to get at least the key heads of terms in writing for the parties to sign before calling it a day. Preparation will ensure that the agreement is appropriate and covers all the essentials. Lawyers occasionally resist drafting an outline settlement agreement before the mediation starts as this is anticipating success and thus inviting failure, but the less superstitious ones will see the prudence in such preparation.

Finally, in connection with anticipating settlement, it is worth thinking about what advice may be needed to ensure that the settlement is workable and will stick. Should an accountant, a tax lawyer, a trust specialist be on hand? Thinking ahead and getting the necessary expertise on standby could make the difference between a deal that sticks and one that unravels as parties try to put a flawed deal into operation.

Chapter 4 in a nutshell

- Choose a mediator with experience over one with specialist knowledge.
- Consider co-mediation in all multi-party cases.
- Minimise the documentation sent to the mediator.
- Keep the mediation team lean and keen.
- Get the decision maker with maximum authority to do the deal.
- Always do a detailed risk analysis. Be confident about your walk-away point.
- Establish needs not wants (for all parties).
- Anticipate the settlement.
- Prepare, prepare, prepare.

CHAPTER FIVE
Presenting at the Mediation

This early stage of the mediation is laying the foundations for successfully negotiating a deal. The mediator, during the opening and exploring stages, will be:

- building rapport (trusting relationships) with everyone, but particularly the parties – if the parties do not trust the mediator, they will not reveal sensitive information or speak frankly about their case, its strengths and its weaknesses
- reopening lines of communication – in most cases, settlement discussions will have reached deadlock and will need an independent influence to reopen them
- identifying the key issues to be resolved in order to achieve a settlement on the day – this may also involve agreeing priorities
- agreeing a timetable for the efficient use of the limited time available and also of the people assembled – this is a unique forum: the people gathered at the mediation will not have met together before and will all be there with the one purpose of resolving the dispute
- trying to understand what lies under the surface of the claims – the parties' needs and drivers and the relative importance of each
- recognising common ground

5.1 Arrival

It is likely that the mediator will be at the venue before anyone else. Rooms need to be checked (soundproofing, natural light, heating, refreshments), seating sorted, toilets located, fire escape routes (and alarm testing) confirmed and refreshment arrangements approved. These are all vital to the smooth running and clear management of the day. The seating arrangements in the main meeting room can be quite a challenge, particularly in multi-party cases. Who sits where, with whom, and, more importantly, opposite whom, is really important if people are going to be encouraged to communicate effectively.

The mediator will try to meet everyone on arrival and take them to their allocated room. This becomes their base for the day, a secure place where files can be stored and parties can relax. Inevitably there will be a lot of files to store. Who knows how the day will progress and what papers will be needed? Usually they remain unused because the amount of detail needed in a mediation is very small – the mediator will be helping parties get out of the detail and more into 'big picture' discussions. But there is always the possibility that a vital document will be needed so all the files are carted to the mediation room. In construction disputes that can mean boxes of files trolleyed in and stored in the corner.

5.2   *Pre-meeting*

The mediator will meet the parties in their room before the initial joint meeting. This is by far the best time for the mediation agreement (see sample in appendix) to be signed. If the agreement is sent to the parties for signature before the day, the result is either lost or forgotten documents or, at best, several documents each with one party's signature. Far better for the mediator to get all signatures on one document and then get it copied and distributed before the opening joint meeting. Whether it is the party or their lawyer who signs is usually of no consequence. Some mediators get everyone to sign. The mediator and assistant will also sign.

Few parties make the best use of this initial private meeting with the mediator. The mediator will use it to reassure the parties (especially if they have not mediated before), explain what they have let themselves in for, encourage the parties to speak in the forthcoming joint session and continue the rapport-building with both lawyers and parties. However, it is an opportunity for the parties to use the mediator

- to request particular information from other parties
- to express concerns (other party's authority, underlying intentions, etc.)
- to demonstrate your reasonableness (in attitude, not case)
- to confirm your wish to work with the mediator in using the opportunity effectively
- to relay any special needs (smoking breaks, access or medical issues, dietary, etc.)
- to give reassurance to worried parties

The mediator is a friend and it is worth building on that. Trust is two-way and getting the mediator to trust you can only be an advantage.

Once the mediator has moved on, it is worth spending a few minutes finally rehearsing your strategy for presenting in the opening joint session. It is easier for the claimants to do this as they are likely to go first whereas the defendants will need to build in some flexibility to effectively respond to the claimants' statements. Be clear about roles, key points to be emphasised and essential aids (flipcharts, PowerPoint, etc.) and documents. There is little point in taking the boxes of files into the joint meeting room; better to take a few essential documents and return for others if it becomes necessary. The thing to remember about the opening joint meeting is to present with confidence, and that means good preparation.

One final point to consider before going into the opening joint meeting. It is worthwhile, if not a necessity, to prepare some questions for your team to ask in the joint session. What else do you need to know to be able to understand their case, their drivers and their needs? More important, what assumptions are you making about the other side's case? Assumptions are flaws in the decision-making process. They need to be turned into fact. So make a rigorous examination of your position and list the assumptions you are making so that you can check them out in the joint meeting. It will be much more difficult to gather that information once you are back in the private rooms.

When you (and the other parties) are ready the mediator will gather everyone in the main meeting room for the initial joint meeting.

5.3   Initial joint meeting

The purpose of the initial joint meeting is

- to set the ground rules (confidentiality, no interruptions, mobiles off during the meetings – there will be lots of breaks to take and follow up messages) – everything is without prejudice if it goes to court (but of course court is unlikely because most mediations settle)
- to outline the mediator's role:
  - to be impartial, with no vested interest in the outcome
  - there not to make a decision or to give an opinion
  - to manage the process efficiently
  - to give the parties the best chance of doing a deal
- to outline the parties' roles:
  - it is their problem, their settlement
  - they are to negotiate in good faith

      ○ there is no settlement without each party saying 'Yes' to the deal
      ○ they have authority to settle (it is very rare for parties to come with full authority to settle and so the mediator will not be surprised if authority is limited – see Chapter 4)
      ○ they can leave at any time – mediation is a voluntary process
      ○ they are not committed to any deal until it is put in writing and signed – this should give everyone the freedom to try solutions and, if they do not work, try something else without the fear of being committed
- to outline the framework of the day (see Chapter 4)
- to check on any time constraints – is anyone needing to leave by a certain time? (not the mediator – s/he should be the last to leave)
- to allow parties to tell their story
- to restore communication (and relationships?)
- to exchange information

Some parties, in particular some lawyers, and (incredibly) some mediators, prefer not to have an initial joint meeting. For all the reasons listed above, this is a mistake. The only time when this might be acceptable is when parties' emotions are so high that they will not sit in the same room together. Even then the mediator may well have several private meetings with the parties, and then a joint meeting once parties have settled into, and trust, the process. One of many reasons why, even in the most emotional of cases, this is appropriate is that parties often demonise the offending party and facing them means that they have to recognise that they do not have horns growing out of their head and that they too are human beings with emotions (and failings). De-demonising parties is one of the important steps in getting people onto a realistic foundation for negotiating a settlement.

There is a fine balance to draw when presenting in the joint meeting. So often the presentations are aggressive, positional and heated, and the result is that the parties leave the opening joint meeting feeling aggrieved and disillusioned. The process is set back and the mediator has to work hard at reassuring the parties and moving the process forward again. However, playing soft is no answer either. A party needs to get their story heard, including the difficult bits, and that may mean being assertive and emotional. But telling the story need not be confrontational. It needs to be clear and informative and aimed at engaging the other party, encouraging them to ask questions and to enter into a discussion. Mediators are taught to 'separate the person from the problem' so that the problem can be seen and examined separately from the emotion. This means recognising and valuing the emotion but seeing the issue as a joint problem: factual and solvable. This underlines the value of having a dry run of the presentation beforehand, not only to ensure the maximum impact but also to reduce nerves and so ensure effective delivery.

There tends to be a pattern to this initial joint meeting. The focus, despite the parties having their say, is usually on the legal arguments, but whilst it is important that the legal context of the dispute is known and provides the backcloth to the dispute, the legal arguments always fall away as the mediation progresses and the commercial arguments take over. It may be that this is changing as lawyers adapt to the relatively new forum of mediation but it is still usual in the greater majority of mediations that lawyers (solicitors and counsel) bring the litigation approach into the mediation arena. Hence legal arguments dominate (but rarely help) in the initial joint meeting.

Case study

A building contract dispute turned on whether a contract actually existed or not. Many months of argument and files of correspondence had not settled the matter and talk was now of getting a judgment on that specific point before pursuing the financial matters. In the mediation the issue fell away in the first three hours. No one in the mediation mentioned contracts again because all the parties wanted was for the financial dispute to be settled and to get on with their businesses. It settled on the day – and still no one is sure whether or not a contract existed!

5.4 Using the joint meeting

Having properly prepared, you are going to use this opening joint meeting to maximum benefit. You are going to

- outline the legal context of the dispute
- tell your story
- hear theirs

I use the term 'hear' rather than 'listen to'. Most people switch off when the other side make their opening statements, mainly because they expect not to agree but also because it may cast doubts onto their own case.

But up to this stage you have probably only heard your truth and this is the time to really get to know the other side's case, their version of the same truth. It is a time to find out if there is anything you have missed, to understand why they feel so strongly about the dispute and to establish their needs and drivers. It is also the opportunity to check out any assumptions that you may have made and to settle any queries. As will be mentioned in Chapter 6, assumptions make decisions very uncertain. One of the purposes of this session should be to turn assumptions into fact. Hence the value of coming into this session with pre-prepared questions so that when you retire to your own room, all (or most) of the assumptions will have been confirmed or converted.

Of course, hearing what the other side say may mean that you need to adjust your own thinking. You must be prepared to change, to adjust your position according to the information you have received. And it is likely that you will hear new information because up to this moment you have probably only heard your own version of the truth. This is where you hear other party's version of the same truth – and why they have a different version to your own. It is important that you hear what they have to say because you need them to say 'Yes' to the eventual deal and that will only happen if you construct a settlement that meets their needs.

Ideally, the joint meeting should be an information exchange and it should not end until you can say 'Yes' to the following questions:

- Do I understand everything about their case?
- Do they understand everything about mine?

Until you can say 'Yes' it is not appropriate for the mediator to end the opening joint session. Some will try. Less experienced mediators will tend to break into private meetings once the opening statements are complete. More experienced mediators will tend to keep the opening joint meeting going until it has ended its usefulness. But the parties can equally be in control of when to break and should have the confidence to resist any attempts by the mediator to end the opening joint session prematurely. This might be the only time in the mediation when everyone is together. This is a flexible process and some mediations have few private meetings; others have only one joint meeting and then the rest of the day is spent in private meetings. So make the best of the initial joint meeting. Don't let it end until its usefulness is exhausted. The thing to remember is that the mediator manages the process but should not control the parties. Keep the session going if it is still useful to you.

Of course, there are exceptions to this. Mediation is a flexible process, and if matters get heated and parties are abusive or the process is being damaged then the mediator would be right to take a break – perhaps to reconvene after the break or to postpone the rest of the joint meeting and reconvene after a series of private meetings. Again, parties should not feel disadvantaged by this and should have the confidence to ask the mediator to reconvene when appropriate so that their information-gathering may be completed.

When the usefulness of the opening joint session is over, it is time to break into private sessions. It is likely to be close to lunchtime and food is usually served in the private rooms. That way the mediator can sample food in each room as s/he shuttles around!

Having said that, there is much to be said for having a communal table. Whilst a mediator justifies having food delivered to the separate rooms on the grounds of keeping everyone working, having the food at a central table can create an informal meeting point for everyone. Eating food together is a communal act that goes back through the history of mankind. It builds rapport and may well cause people to communicate more easily than in the meeting rooms. At least one mediator encourages this and sets a rule that no one should discuss the case whilst they are eating. One other well-known mediator has the practice of dining with the parties (together) before a mediation. I am not too sure who pays the bill! Certainly, if the mediation is at a hotel and many of the parties are there overnight, I would invite them to dine with me on the understanding that no one discusses the case. Such mediations always start the next day in a less confrontational way.

5.5   Exploring stage

Once into private sessions the mediator will want to explore, in confidence, the parties' positions. Well-trained mediators know that the foundations of a successful settlement are laid in the exploring stage of a mediation by

- building a more complete picture of the dispute
- getting to know the *needs* of the parties rather than their demands
- consolidating the rapport so that parties will speak openly and frankly
- helping parties to move from the history of the dispute to the future – from looking back to looking forward; from the misery and negativity of the past to the freedom and opportunities of the future

The past cannot be changed but the future is in the parties' control (control that is not possible in dispute processes outside negotiation and mediation).

The term 'needs not wants' has been mentioned several times already. What is it that parties really need to be able to do a deal? Often it is very different from what they claim and just does not come out in litigation or arbitration.

Case study

A young plastering sub-contractor was suing a builder for outstanding payment of invoices. The contractor counterclaimed that the work was not up to standard and he had had to pay another contractor to put the work right. The sub-contractor's lawyer was asserting that payment was due in full because the sub-contractor had not been informed of any defective work or given the opportunity to rectify if defects existed. When the mediator asked the sub-contractor what he really needed to be able to do a deal, he said, 'I just want to go home and spend time with my little children.' The case had so dominated his life that every moment was spent thinking or dealing with the dispute. He struck a deal – not what he wanted, but it gave him what he needed.

A well-known diagram (called the PIN diagram), devised by Andrew Acland,[1] which we use in our mediator training course, illustrates this (Figure 5.1). It is like two icebergs. What people claim is what you see above the waterline (their 'position'); the two positions are incompatible (hence there is a dispute). Below that – and this is what the exploring stage should expose – are the parties' real interests. What do they really want to achieve? What are their drivers, what is really important to them? And below that are their needs and fears. What do they need to be able to achieve a settlement of their dispute? As the diagram suggests, the parties needs are usually much closer than would appear from their positions and often there is an overlap. It is at this level that the common ground can be established and the foundations of a settlement can be cast.

---

[1] Andrew Acland, *Resolving Disputes without Going to Court.*

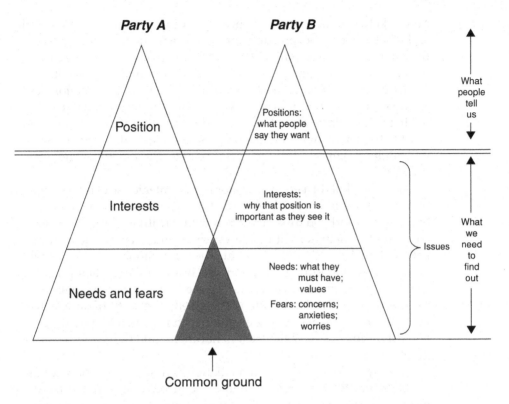

Figure 5.1   PIN. 'iceberg' model.
Courtesy of Andrew Ackland.

5.6   *Giving and receiving information*

The mediator is in a unique position of gathering information from both/all parties in the mediation. S/he has an insight into parties' needs, their attitudes and responses, and their financial and risk calculations. No one else is in that position and it is something that should be used to advantage by the parties. Much of the time parties need reassurance that the other side is there in good faith. Information is not going to be passed and offers made if there is a suspicion that they are there in bad faith – for example, on an information-gathering expedition with no intention of settling. So the mediator is in a position to give that assurance and to encourage the parties to be open and frank about their position in return.

Confidentiality is the first rule of mediation. The mediator may say that everything said in private session will be treated as confidential.

More likely, the mediator will check what information is confidential and what is not. Some, probably more experienced, mediators will confirm specific information and leave the rest open. It is therefore up to the parties to be specific as to what is confidential and what is not. And whilst the temptation is to play safe and keep most information confidential, the question to ask is: 'Why should it be confidential?' The mediator is in the best position to decide what information to pass and what to hold: what will help the progress towards settlement and what will hinder progress; when the time is right to say something and when it is not.

One of the skills of a good mediator is the strategic use of information. S/he needs maximum flexibility to do this effectively. If the mediator has built rapport and you trust her or him with sensitive information, don't constrain the mediator with unnecessary confidentiality. The mediator is there to help parties get to a deal. The information s/he carries will be used with that purpose in mind. Of course, if the mediator has not built rapport then the trust will not be there to give parties confidence to pass on sensitive information. The essential quality of an effective mediator, more than anything else, is the ability to build a trusting relationship quickly. So if the trust is not there, don't give the sensitive information.

It is easy to be impatient in a mediation and want to get on with the negotiation. Mediators are sometimes 'seduced' into skipping the exploration because parties say, 'We know all about their case, so stop the touchy-feely stuff and get on with the figures.' It takes a confident, if not brave, mediator to resist that and stick with exploration. But many experienced mediators will tell stories of when they complied with the suggestion to go straight to the figures and then the mediation deadlocked. They had to bring parties right back to exploring needs and drivers to get the mediation on track again. The fact is that parties have probably already tried negotiating a settlement and reached deadlock (otherwise they would not need a mediator). Mediation offers a different way that has been tried and tested over many years and thousands of mediations. So trust the process and be patient! Exploration is important and creates the foundation for the best settlement.

5.7 Idle time

Having mentioned that the mediator shuttles between parties' rooms and holds private meetings, it is necessary to highlight the one problem with mediations. The idle time – time when the mediator is with the other party or parties and you are in your room waiting patiently for the mediator's return. It can be a time when the process is set back, when people disengage because they are bored and when they start to think about things that they could be doing if they were back at the office. Idle time is inevitable in mediations so it is prudent to plan for it: not, as some mediators would suggest, by bringing a book or the newspaper – that encourages disengagement; more by using the time creatively, both physically and mentally. The mediator will be very aware of the problem and should manage the idle time effectively by

- setting a task – something for you to do whilst the mediator is with the other party
- keeping the private meetings relatively short – this means that the idle time is limited and also creates a momentum to the process; the mediator does not have to cover every issue in every meeting, and indeed it is much better to leave some issues outstanding so that a party can prepare to discuss other issues when they next meet
- encouraging parties to get some exercise and fresh air so that everyone is still as fresh and active in the late afternoon as when they started
- creating working groups, perhaps experts or lawyers together – this at least keeps some people working but does leave other (though fewer) people idle; if the mediator has an assistant they may sit in and perhaps chair a working group, if only to keep the discussions on task and ensure the group is being productive – of course, in a co-mediation the mediators are likely to split and be in separate meetings anyway

The best way of avoiding idle time is to keep everyone together and so, whilst private meetings are essential for parties to discuss confidential matters, with or without the mediator, joint meetings are by far the most efficient and will ensure that people remain engaged in the process. They also mean that everyone hears the discussions first hand and, even if people do not participate, they get the same story. So often when people return from working groups they filter their reporting back and each side receives different stories, or different emphasis.

5.8   *Others' shoes*

I mentioned earlier that different people see the same thing differently, owing to a whole lot of reasons, and this does not necessarily mean that they are any more right or wrong, just different. One of the things that mediators will do at this stage of the mediation is to encourage parties to step into the other side's shoes (or look at the issues through their eyes). The deal is only possible if both/all sides say 'Yes', so envisioning the issues from the other side's position will help in crafting a settlement that meets both parties' needs. Adopting 'their' position can be very revealing and some mediators may go as far as getting a party, in private, to physically move position, sitting in the other party's chair, or just adopt a position that the other side had earlier. This may be the key to understanding why they see the issues differently, and that understanding will help the crafting of a solution – the best solution – to which all can sign up.

5.9   Non-financials

It is often stated that mediation 'enlarges the pie'. In other words, the settlement options are often, if not always, greater than that which a court (or arbitrator/adjudicator) can offer. The settlement can include anything that is legal. And non-financials can make the difference between settlement and no deal.

Case study

A dispute over payment of some building works at a football club was deadlocked. The building company was a small operation and the work had been done at a low price because the owner was an ardent supporter of the team. However, the club had limited funds available, insufficient to settle the total bill if it were proven to be reasonable. The builder feared that the dispute would give him a bad name and that he would no longer be welcome at the club. The case settled when the club paid a small sum and gave the builder a five-year season ticket.

Whether or not the mediator raises the matter of non-financials, it should be raised and discussed by the party. Non-financials (that is, items beyond the money sum) such as

- deferred or stage payment (or accelerated payment)
- future business/discounts

- services in lieu of cash
- goods/property in part payment
- merger/takeover/joint venture

and so on. Be creative – the possibilities are endless. This might be a case for using the flipchart and doing some brainstorming. There is nothing to lose and may be everything to gain.

Case study

A consultant's insurance denied liability on a claim which would have left the consultant bankrupt. The eventual settlement involved a small payment together with an agreed number of hours' free consultancy to be called in over a three-year period. The consultant survived and continued to work with the client long after the agreed hours had expired.

Better solutions are usually created in mediation because it is free from the constraints that bind the court/arbitrator/adjudicator.

Chapter 5 in a nutshell

- Let the other party tell their story. This is their 'day in court'.
- Be informative, not confrontational. The other side must know your version of the truth.
- Hear what the other side have to say. There may be some useful insights. Even if there are not, they will feel you have listened and understood what they have said, and their attitude will soften as a result.
- Prepare for and ask questions. This is the opportunity to gather information that may affect your case.
- Turn assumptions into fact.
- Do not allow the joint meeting to end until you can say 'Yes' to the questions 'Do I understand everything about their case?' and 'Do they understand everything about mine?'
- The past cannot be changed but in mediation the future is in the parties' hands.
- Don't clothe everything in confidentiality. Be specific about what is confidential and give the mediator maximum flexibility.

- Prepare for the idle time and plan to use it well.
- Stand in the other party's shoes. Understanding their needs is vital to crafting a solution that is best for everyone.
- Think creatively. Settlements do not have to involve money alone.

CHAPTER SIX
Negotiating at the Mediation

This is what you are here for – to negotiate a deal. With the help of a neutral third party – the mediator. One question that needs to be asked is: 'How much do we use the mediator and how much do we negotiate direct?' My feeling is that experienced negotiators like to negotiate their own deals. Seeing the other side's reaction to statements or offers can help inform what to do next. On the other hand, using the mediator as a sounding board can help to pitch offers in a strategically efficient zone. Whatever the decision, there is much that can be done to ensure that the best is made of the opportunity presented by mediation. And if you do make the best of the opportunity, it will put you to advantage because most mediators experience pretty awful negotiation 'skills' by parties, and their lawyers, in mediation – so much so that I want to hand them my card at the end of a mediation and say 'We run a good negotiation skills course that will help you next time'!

6.1 Negotiation zones

Pepperdine University[1] suggests that a typical negotiation zone graph would be as shown in Figure 6.1. It is rare indeed for offers to start in the likely zone of agreement, but the closer offers are to this zone, the quicker and more good-natured the negotiations are likely to be. The mediator should spend time getting parties into a position whereby, when offers are tabled, they send the right message and encourage a positive response. If not in the zone of agreement, opening offers in the reasonable zone dictate the negotiation parameters. They are reasonable enough for the receiving party to be able to predict the area of likely agreement. The offering party will not have much room to manoeuvre but the message is clear that 'I am being reasonable, and you know it'.

Offers in the credible zone are likely to be received as 'acceptably unreasonable'. They are likely to be seen as positional but with a hint of

---

[1] www.pepperdine.edu.

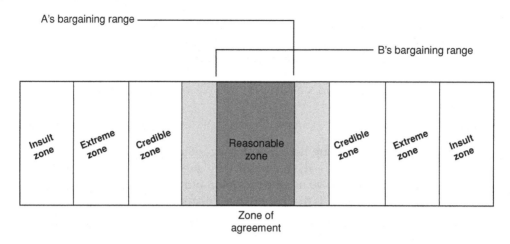

Figure 6.1 The zone of agreement.

cooperation. The respondent is the one who is likely to set the parameters of negotiation.

Offers in the extreme zone do not set the parameters for negotiation, may be received as an insult and usually result in the offering party having to concede quickly. It therefore affects credibility and usually puts the offeror to a disadvantage.

Offers in the insult zone are likely to be received as so unreasonable that the receiver may refuse to negotiate. Some positional negotiators may take pleasure in pitching offers in this zone but it invariably puts them to disadvantage.

6.2   Reviewing

Following the opening joint meeting parties will retreat to their rooms and take stock of the situation. If the joint meeting has been used well – and it should have been, if it was well prepared and planned beforehand – there will be some useful information to analyse and discuss.

One area of the risk analysis is the cost of not settling in the mediation. Not just the legal costs of preparing for and attending the trial/hearing – the management costs can be huge and are almost always underestimated; the emotional costs are less easy to quantify, but few people enjoy the experience of standing in a witness box and being cross-examined by someone whose aim is to discredit them and to undermine their version of the truth. There is a typical risk analysis schedule in the appendix which attempts to quantify the true cost of taking a case to trial or arbitration. Most mediators will touch on this issue during the mediation, some even at the very start. I tend to feel that it is rather a blunt instrument, in danger of being seen as a form of blackmail to get the parties to settle, and so I leave it as the last resort in any negotiation. Having said that, the cost of going to trial or arbitration can be one of the most powerful reasons for parties settling in the mediation. The estimated legal costs can frequently exceed the value of the claim, leading to a situation where no one wins.

Reviewing the risk analysis should be a constant process throughout the mediation: assessing the information and amending the financial calculations (see Chapter 4 regarding risk analysis). The BATNA (best alternative to a negotiated agreement) needs to be reviewed and updated so that the 'walk-away' point is clear. This provides the foundation for confident negotiation, having a logical and sensible bottom line.

6.3 Bottom lines

A sensible mediator will avoid asking a party for their bottom line, their 'walk-away' point. The reason for this is that bottom lines change as the mediation progresses – or should do. If the mediation is being used to best advantage, the gathering of information will inform parties of how their bottom line may be adjusted. Therefore a bottom line at an early stage is unlikely to be the bottom line in the final negotiations. If a mediator asks a party for their bottom line it is quite likely that s/he will have to manage a 'face' issue later on. Having stated a figure to the mediator, a party may well find it difficult to move. So, whilst a sensible negotiator will always know where the bottom line is, and will constantly review it, it will be confidential information rarely shared with the mediator.

A potential concern that a party may have in letting the mediator know a bottom line is that the mediator will know the figure at which they are prepared to do the deal. The worry may be that it will be more difficult to get a better deal than the bottom line because the mediator will have the figure in his/her mind. However, it would be much better to avoid giving any figure to the mediator rather than give a false one and be potentially faced with admitting that it was untrue later. Such a situation will not only undermine a party's credibility but also challenge the trust that should have been built with the mediator.

6.4   *Negotiation strategy*

Proper preparation means that the negotiation strategy will have been fully worked out beforehand. This does not mean that it is unalterable, but it does mean that a party will go into the negotiation stage with a clear purpose and focus. The only problem then is if the other side's strategy is incompatible and you are faced with the challenge of changing or persevering with the one you had planned. The best negotiations are when both parties cooperate and make sure that the best deal – one that uses up all the potential settlement options – is achieved. Some people see principled negotiators as being 'weak' because they won't enter the 'gladiators' arena' and fight toe to toe. The fight for them is what really matters. Men are not meant to be nice to each other. Yet the best way to counter such mistaken attitudes is to be confident about principled negotiation and not to join in the positional fray. Having planned your negotiation strategy beforehand, stick to it – always assuming, of course, that your strategy is to be principled in negotiation. If your strategy is to be positional, be prepared to abandon it! And read *Getting to Yes!*[2]

6.5   *Incompatible styles*

Having extolled the virtues of principled negotiation, it has to be admitted that unfortunately most parties (or, rather, their lawyers) negotiate from positions, not principles. They stake their claim at its most extreme with the expectation of moving, gradually, to a deal. The theory is that to give little, slowly, gives you an advantage. But no mediator enjoys parties 'salami-slicing' by shaving small amounts off the claim/offer, because it is not only a waste of time but is very inefficient and can make parties very cross. It becomes a ritual. Experienced mediators know that the best negotiation strategy in a mediation is to foster cooperation from the start, avoid offers in the insult and extreme zones, and send a message that a party is genuinely here to negotiate a settlement. Some positional negotiators may respond by saying that is a reason for pitching the first offer in the insult/extreme zone, sending a message that they may be prepared to pay something but it won't be much and, whatever it is, the other side are going to have to work hard to get it.

---

[2] Roger Fisher and William Ury, *Getting to Yes*.

But in reality it is a false premise. Parties are there to do a deal, and the short, and cooperative, way to do so is to pitch a figure in, or close to, the probable zone of agreement. If the mediator does his/her job, offers will not be tabled until there is a willingness to pitch sensible figures.

If a principled negotiator (that is, someone who focuses on the needs of the parties) has a positional negotiator on the other side, it can be a challenge to resist reverting to positional tactics. It is a matter of holding nerve and holding on to the belief that a principled approach results in much richer settlements.

Principled negotiation ensures that the best deals are obtained and that nothing is left outside the deal. So make the offer (preferably the first one) and hold the nerve, whatever the response. You are in the moral high ground – let them come to you!

Case study

In a professional negligence claim of £1 million, with the insurer present, the opening offer by the defendant was £50,000. The claimant responded with a counter offer, with the preface of 'for the purposes of this mediation we are prepared to settle our claim at £450,000 plus costs. Our reasons are as follows . . .', and they set out a logic based upon the lawyer's estimate of success in court under the various headings of claim. The mediator urged the insurer defendant to reciprocate by raising their offer to a level that might be tempting to the claimant. However, the defendant responded by raising their offer to £100,000 inclusive of costs. Disappointed, and not a little angry, the claimant confided in the mediator that they felt they had misjudged the negotiation and been disadvantaged by making a reasonable opening offer. Right or wrong, there was no going back and it was agreed that the mediator should go to the other party with the message that the claimant would stick to the opening figure until the defendants treated the claim seriously. They were prepared to leave the mediation if the defendant persisted in taking 'such an extreme position'. The mediator spent a long time with the defendant discussing their options and eventually a settlement offer of £250,000 came back, with the mediator given flexibility over costs. Eventually a deal was struck at £350,000 plus costs, but the mediation was seriously jeopardised by the defendant's tactics.

6.6   First offers

Most principled deals settle in three moves – initial, subsequent and final offer. Positional negotiations take much longer, tend to cause bad feeling and generally end with grudging and inefficient deals.

I always feel that the first offer not only sends a message but, if it is sensible, sets the zone of negotiation. Offers that follow are gauged to the first offer. A reasoned and reasonable offer is very compelling. It is likely to reassure the other party that a party is negotiating in good faith and will encourage the other side to respond likewise. An opening offer in the reasonable zone sends the invitation to be cooperative. If the other side persist in positional offers, then the cooperative message needs repeating, perhaps reinforced by adjusting a little or just repeating the offer already tabled. If they still do not respond in like terms then certainly stick. Keep a little in hand for the final offer but let them work to get you to move.

One reason why a party may be reluctant to make an opening offer may be the fear of giving a better deal to the other side, but this all comes down to preparation. If a party has prepared properly, and gathered and checked information in the opening joint meeting, they should be confident in their valuation of the case and therefore the level of sums to be negotiated. Making the opening offer, particularly if it is accompanied by a logical explanation, means that they set the stage for the subsequent discussions.

6.7   Offer logic

Offers, particularly ones outside the likely zone of settlement, have much more credibility if they have an explanation – for example, percentages of figures under the various claim headings. It deflates the likely response that such offers are not serious and gives those receiving the offer information to work on in reply. Of course, that is the potential disadvantage of giving a logic to an offer – it provides the other side with something to fire at. It also gives the mediator an insight into where the differences – and the common ground – might be. Again it all comes down to preparation. If a party is confident about their negotiation strategy, having their logic 'fired at' will not cause concern – and it might even produce some useful information because the other side are likely to provide a logic to their response. Making the first offer gives an advantage – it sets the agenda for the subsequent discussions. However, whether it be the first or subsequent offer, it is always best to give the logic before the figure. People get hooked onto the figures, and if the logic follows it is rarely heard.

Of course, the final offer need have no logic. It is usually a global figure that is judged to be the best to do the deal. I often say, 'There is a figure out there that both of you can say "Yes" to. The challenge is to get to it by the quickest and least painful route.'

It is worth considering how much of this happens through the mediator and how much is face to face. Much will depend on how confident a party is with the mediator. Will the right message be relayed? Will s/he explain the logic clearly? Would it be safer/more efficient for it to be done face to face? If so, by whom and to whom?

My feeling is that most effective negotiators like to see the other side's reaction first hand. They can see the eyes, hear the tone and inflection, read the body language and gauge reactions. Most effective mediators will understand this and offer the negotiators the choice, particularly when the negotiations near settlement (or breakdown).

6.8   *Getting into deadlock*

It is a common saying on training courses that most mediations reach a point of despair. Of course, most negotiations are in deadlock already when they get to mediation. Part of the mediator's job is to reopen lines of communication and get the negotiations under way again. But it is rarely easy and deadlock may well occur again – a point where everyone, including the mediator, is banging their head on the wall and wondering what on earth to do next. The very fact that everyone is feeling the same way is probably why so many mediations then tumble to settlement. It is as if this point has to be experienced before the final steps are taken towards settlement.

The mediator will have several strategies available to break the deadlock, for example:

- Taking a break – going for a walk, getting some fresh air, is often the key to breaking an impasse.
- Better still, the key decision makers going for a walk together.
- 'Chunking' down/'chunking' up – breaking big issues down into smaller ones or, more likely, discarding the detail and looking at the big picture.
- Changing the groupings – putting just the lawyers together, or setting up working groups on specific issues.

- Reviewing progress – often new energy comes from realising how far matters have moved in the mediation.
- Brainstorming – getting people to think creatively, preferably together.
- Deadline – the mediator setting a deadline for the impasse being broken or the mediation ceases: this is not a solution that I favour, as the last thing I want is to suggest an end to the mediation – I would rather agree an action timetable whereby parties have specific tasks to complete by a specific time.

One danger of deadlocking is the fear of losing face in agreeing an outlet. The mediator needs to find reasons for parties to move without it appearing to be capitulation. Face is an issue, particularly in the negotiating and concluding stages of mediation. People need to do deals with dignity.

6.9   Pain–pain

The sad fact is that most commercial mediations are about money and rarely settle at the amount each party wanted to give/receive. Most deals are done with a bit of pain. In the early days of mediation, the phrase most commonly used was 'win–win situation'. It originated from the book *Getting to Yes,* which propounded principled negotiation. Win–win situations arose because the settlement pie was bigger and parties cooperated in achieving a settlement that met both/all their needs. The important thing though is that parties are prepared to take that extra painful step to achieve a settlement, provided the other side are suffering as well. Shared pain is more acceptable than pain that is one-sided. Deals with dignity again.

The mediator has a key role to play here. Truthfully relaying the fact that the other side are equally trying hard, and suffering, to achieve a settlement, without breaking any confidences, can be very encouraging. Of course, if this is not truthful, the mediator's credibility will be blown. Honesty and openness are key qualities of any mediator. Anyway, if one side is suffering and the other side not, the mediator should be working harder on the latter!

6.10   Keeping options open

One of the key principles of mediation is that no one is committed to anything until the deal is put in writing and signed by the parties. This gives everyone the freedom to try options for settlement and discard them if they do not work. Similarly, it means that decisions on individual pieces of the jigsaw do not have to be agreed until the whole jigsaw is complete. Sorting one issue and then parking it and moving on to the next issue is a common approach. Parties usually need to see how everything fits together before finally committing themselves to a deal. Most often, it is the final sum that is the main interest (to all parties), and the small pieces are just steps on the way.

Case study

A four-party construction mediation involved an adjoining owner claim, the developer of an adjoining site, a structural engineer and main contractor, and concerned the subsidence of an adjoining property that had been inadequately shored during excavation works on the adjoining site. The issues were calculation of damages, negligence, appropriate mitigation and consequential losses arising from delays to the adjoining works. Insurers were present. There were in effect three mediations:

1. level of damages paid to the adjoining owner (developer/adjoining owner)
2. who paid and what proportion (structural engineer/contractor)
3. amount of delay and loss and expense (developer/contractor/structural engineer)

The first mediation achieved a settlement subject to the second being resolved (in other words, the amount of damages was agreed subject to the insurers agreeing apportionment). The apportionment was agreed subject to the third mediation being resolved because the structural engineer's insurers were also potentially liable for a similar proportion of any payment justified by the contractor. In the final stages of the mediation all three settlements happened in the space of half an hour once the various sums had been agreed. In the end the developer agreed not to deduct liquidated and ascertained damages and the contractor dropped his loss and expense claim, which in turn enabled the structural engineer's insurer to agree a higher percentage of the damages to the adjoining neighbour. Everyone smiled, signed the settlement agreements and had a glass of wine.

6.11 *Non-financials*

As mentioned earlier, deals do not have to be money-only – although most settlements in commercial mediations are about money. Any deal is possible in mediation (provided it is legal). Mediation allows anything to be put into the settlement pot, even other disputes. Other dispute resolution processes cannot do that – they are law-based and usually either prevent someone from doing something or award damages. So the mediator will (or should) encourage parties to think creatively about getting the best deal possible. What else besides a sum of money could help get the best deal? Whatever both/all parties can say 'Yes' to makes the deal. It is worth preparing for this moment and coming to the mediation with some (sensible) suggestions that could be the key to obtaining a deal.

6.12 *And finally*

An excellent question for a principled negotiator to ask is 'What can I give them to enable them to give me what I need?' Sometimes a small concession that is inexpensive to one side may be highly valued by the other and can be an incentive for them to reciprocate with a concession of their own. When negotiating, it is worth looking for differences in values to identify where gains can be most easily made. For example, an apology/expression of regret/acknowledgement of distress can have a powerful effect. Even saying something like 'We were all so pleased when we started off – it is so sad that we have ended up here' gives a message that is free from blame and is a reminder of good times past. Mediation frames disputes as joint problems, offering joint solutions. They are in this together!

Chapter 6 in a nutshell

- Review constantly.
- First offers usually dictate the zone of negotiation.
- Offers with logic get a more positive response.
- Principled negotiators get better deals.
- Expect pain. No one said this would be easy!
- Be creative. Mediation offers limitless possibilities.
- Hold your nerve!

CHAPTER SEVEN
Concluding the Mediation

The term 'concluding' is deliberate. Most mediations settle on the day – but a few do not. Most of those that do not, settle very soon afterwards and the mediator, understandably, claims those as being settled because of the mediation. Quite how far that can be taken by a mediator is not clear. One or two weeks after the mediation day would definitely be a justified claim by the mediator. Two to four weeks, maybe. After the judge's/arbitrator's decision, definitely not (although it has been known for a mediator to claim this as a 'settled' mediation)!

7.1   Finality

One of the key benefits of mediation is the fact that it offers certainty in outcome. Parties can leave a settled mediation with the confidence that the dispute is over and tomorrow is another (brighter) day – whereas relying on the judge/arbitrator to decide has a great deal of uncertainty and not necessarily any degree of finality. Once the parties have agreed a settlement in mediation and it is put in writing and signed, the matter is over. The lawyers have a satisfied client and the parties can get on with wealth-generation rather than wasting valuable management time on defending/pursuing a claim. Most people like certainty. Mediation can give it.

7.2   Deals with dignity

'Deals with dignity' is a term that I have used elsewhere and often mention in mediations. Most parties understand the need to reach an agreement that leaves the other side with some feeling of justice, or fairness. There needs to be a feeling of 'honours even' with no side being the outright winner – or loser. Deals with dignity mean that parties can leave with heads held high, face saved and needs met (at least in part). The mediator will try hard to ensure that the ultimate settlement will meet this criteria. However, some negotiators just want to win. It may have nothing to do with a party's needs, other than to have an ego fed. And sometimes they succeed and the other party leaves the mediation feeling aggrieved and 'shafted'. Try though the mediator may to dissuade a party from taking such a course, there is little that even an experienced mediator can do in these circumstances – in the end it is the party's responsibility and they can walk away from such a deal if they do feel strongly about it.

Case study

In a construction dispute the developer prided himself on being a good negotiator. He had driven a hard bargain and the parties were stuck on being only £10,000 apart on a deal that exceeded £250,000. Both threatened to walk away: the claimant because he felt he had already given away too much; the developer because he sensed he could win a better deal by holding out. In the end the mediator said in exasperation, 'Oh for goodness sake, split the difference and let's all go home.' In private the claimant told the mediator that he would agree. In private the developer offered a further £2,500, despite his lawyer's advice to the contrary. The mediator sensed that this would be the last straw for the claimant and that he really would walk away. He spoke privately with the developer's lawyer, who eventually persuaded her client to offer the £5,000 and so do the deal. The developer agreed and the deal was done. Whilst the lawyers were writing up the settlement agreement, the developer said to the mediator, 'I enjoyed myself. I am a good negotiator.' Right or wrong, the mediator replied, 'No, you just like winning.'

We live in a small world where even the most unlikely paths cross again. Construction is a big industry but one built upon continuing relationships because it operates in reasonably contained communities. Good relationships mean good business. Being 'shafted' destroys good relationships and earns reputations that challenge trust and reduce the opportunities for working effectively together. Deals with dignity preserve and even rebuild relationships which may well mean good business in the future.

7.3 Part-deals

Just occasionally parties are unable to reach a full settlement but are able to reach agreement on certain issues. A deal in mediation may also include a strategy that will lead to settlement on the other issues. Similarly, in multi-party cases agreement may be reached between some parties but not all. In mediation anything is possible.

Case study

In a four-party construction mediation involving a university client, architects, main contractor and mechanical and electrical sub-contractor, the main issue could not be resolved because of an insurer's unrealistic expectations. However, the main contractor and sub-contractor reached an agreement during the mediation, thus simplifying the dispute and removing some of the issues disputed.

Although mediations within mediations are not uncommon in multi-party cases, the majority of cases are two-party and often single-issue. Hence the high settlement rate. The more complicated a case, the more parties there are and the more issues to be resolved, the more difficult is settlement. But settlements still happen in most cases – the fact that a case is complex does not mean that a judge or arbitrator is better suited to resolve it!

7.4   No deal

Sometimes, rarely, deals are not possible in mediation. There could be a variety of reasons:

- One party, or both parties, has not prepared properly.
- New issues have arisen that require further investigation.
- Parties are mediating too early. Although most mediators would say 'mediate as early as possible' to avoid the escalation of costs, as mentioned earlier, parties need to be sure about what issues are to be mediated and their respective positions on them.
- Parties' risk assessment is completely incompatible. This will affect a party's expectations and even an experienced mediator may not be able to manage those expectations sufficiently for a realistic settlement to be possible.
- Authority to settle is insufficient or the real decision maker is not present.
- A point of law is still seen to be crucial and outweighs the commercial logic to settle.

- The costs already incurred are so large that they exceed the claim/likely settlement (although that situation can only get worse if the matter continues to court/arbitration). Sometimes a party is trapped by the level of costs and it is too late to 'step off the tread-mill'. No lawyer can take pleasure from this situation, for no one can win in the end, but it is usually the lawyer who decides when to take a case to mediation and the guide should be as soon as possible, if only to avoid costs becoming the overriding issue.
- A party's needs are not being met, or are still not known. This rather reflects on the mediator, who should have ensured that parties' needs are known and that any settlement reflects those needs.
- Inability to pay. Of course, inability to pay in a mediation can only get worse if the matter continues to court or arbitration.
- Emotional issues, not least that a party has invested so much in the dispute that letting go may be very difficult. The paying party may have to accept that settlement in these circumstances often has a premium.
- 'Matters of principle' – these may fall under the previous category. The last thing that a mediator wants to hear is a party saying, 'This is a matter of principle.' It usually means that all common sense has been lost and no amount of reasoning, particularly by the party's own lawyer, will sway the party. This becomes a matter of face and mediators may not always see it coming or be able to deflect the issue.

Of course, most mediations settle, so this situation is rare. Even then, it should be possible to agree a strategy to overcome the blockage and so maintain the momentum towards settlement. For example:

- Agree an action plan.
- Postpone the mediation/agree a further day/half day.
- Get an opinion on a particular issue or legal point. Early neutral evaluation may be appropriate (a judge's forecast of the likely outcome if the matter were to go to trial), or a jointly appointed expert's report.

Case study

A multi-party international construction dispute settled in a two-day mediation on all points except the relevant interest to be applied to the settlement sum, as the dispute had been rumbling on for nearly four years. It was agreed that, because at least one party had

---

time constraints, the lawyers would agree a joint letter and request a jointly agreed bank to specify the relevant interest rate(s). All the other issues were agreed and documented, subject to the resolution of this one outstanding item.

Unfortunately, the three lawyers could not agree a joint letter, wrote their own letters to three different banks and obtained three different rates of interest. The matter took several months to resolve and for the original agreement to be effective. By which time of course, the claimant wanted even more interest!

The lesson for the mediator was not to let parties leave, even if they did have a plane to catch, before all the issues were resolved.

7.5 Writing the settlement agreement

Most settlement agreements from a mediation are simple. The best are brief, record the key points and terms and avoid complicated language and clauses. The mediator, or a party's lawyer, may have a typical agreement wording to use as a template. If no court proceedings have started it need only be a straightforward Heads of Agreement. If action has started the agreement will be in the form of a Consent Order or Tomlin Order (the latter keeping the actual terms of the settlement confidential as an attached schedule to the court order).

Understandably, mediators are reluctant to write the settlement agreement. Oversee it – yes. Offer ideas and wording – definitely. But not write it. After all, most commercial mediations have lawyers present and it is they who have that responsibility. It can be a frustrating process, and a lawyer-mediator may be tempted to act as a neutral scribe, but the drafting should be by others. It is their problem and their deal. It is inevitable that writing up the deal takes some time. Every word becomes important and each lawyer must ensure that his/her client's interests are protected. But it is also an opportunity for the mediator to be creative.

As a non-lawyer mediator I am happily unqualified to write a settlement agreement, so my attention is taken with filling the time, which is commonly a couple of hours, to the parties' benefit. Once the deal is done there is often a sense of relief, even achievement, and it is an opportunity for the parties to meet again, free of the dispute. This is easier if the decision makers themselves have struck the deal, for the relationship will normally have already been restored. If the mediation takes place in a lawyer's office there is invariably a wine store that can be raided so that the parties can have a drink and chat together in the main room. If the mediation is in a hotel there is usually a bar where people can meet whilst the lawyers work. Most of my mediations end this way – and it is another demonstration of the power of mediation. Parties just would not have a drink together if they had slugged it out in court or an arbitration hearing; usually they never want to meet again. In mediation it is different, and fractured relationships can be restored. Indeed new business is often discussed.

Case study

A factoring company was in dispute with an engineering company who had, after a long association, changed factors and allegedly broken a contract. Much of the cause of the dispute was a breakdown in communication between the two companies and ineffective delegation of responsibilities. The two principals, who had agreed the original contract, no longer had direct contact – until they attended the mediation. The dispute settled in mediation. Over a drink whilst the lawyers were writing up the settlement agreement, the principals agreed a new factoring contract for a subsidiary of the engineering company – and to meet regularly to avoid a repeat of the dispute.

7.6   What can go wrong?

The mediator will be asking questions of the lawyers writing up the settlement agreement. For example:

- What has been learned from this that will help to avoid the same situation arising in the future? If there is no continuing relationship this may not be relevant to the agreement, but it will still be relevant to the party.
- Date for payment.
- Default provisions; interest for late payment. Mediator to adjudicate on any disagreement arising from the settlement. The mediator is the custodian of the spirit of the deal.
- Who pays legal costs? Usually any deal will include costs or be subject to taxation. But assumptions may have been made and the settlement agreement must be specific.

- What are the tax implications (whether VAT, income tax, inheritance or charity, and so on)?

> ## Case study
>
> A health authority agreed a settlement in the mediation of a delays and loss and expense claim on a hospital project. As the settlement was being written up the mediator asked, 'What is the VAT position?' In the weariness of the long negotiations, it had been forgotten that the health authority could not reclaim VAT. They had assumed the agreed deal was VAT-inclusive; the contractor had assumed it was subject to VAT. The difference was potentially a deal-breaker and there was another round of lengthy negotiations before the deal was agreed and the settlement signed.

If parties have prepared properly, they will have experts on standby if needed, particularly if income tax or charity law are involved.

People are often weary at this stage of the mediation, especially if it has gone on into the night. Even lawyers will not be at their best at this time and it may be unreasonable to expect a clear and watertight agreement to be produced even in a couple of hours. But it is important that the key points of the deal are recorded and signed by the parties before they leave, even if the lawyers have to write up a more detailed agreement the next day. Again, the mediator is the custodian of the deal and is a reference point if difficulties occur during a more detailed writing up of a contract or agreement.

7.7 Cooling-off period

One of the mediator's responsibilities is to make sure the eventual deal will stick. This involves serious reality-testing during the mediation and continuing to the end. If the mediator is at all concerned about the sustainability of the agreement s/he should, in private, test the parties to be absolutely sure that the deal is workable. In particular, if a party appears to have been bullied into a deal, or if a party is unrepresented and so has no legal advice, the mediator might suggest a cooling-off period to give the parties time to consider the deal and to take advice if necessary. Some may say that this is a dangerous thing to do as it may cause the deal to unravel; after all, one of the principles of mediation is to get the parties committed on the day. But the mediator's responsibility for a deal that will stick overrides any worry there may be that parties may have second thoughts if given a cooling-off period. If the deal is likely to unravel tomorrow then it should not be a deal that will stick today.

Settlements could be subject to ratification within a day or two, or firm unless a party withdraws within an agreed period. In the case of public bodies it is more likely that a period of two weeks or more will be required to get a committee or authorised officer to ratify the deal. The deal is whatever the parties agree, and a cooling-off period or time to obtain final approval is not uncommon.

7.8   *Mediator recommendation*

In the past, some mediation agreements have had a provision that the mediator may, if so requested by all parties (and if s/he so agrees), make a non-binding recommendation for settlement. It is a dangerous clause and one that most providers have removed and most mediators resist. Of course, the advantage is that the mediator is in the best position to know what the parties need to achieve a settlement, so any recommendation is likely to take this into account. However, the dangers are as follows:

- The mediator will use, and may reveal, confidential information given by one or all parties. With this clause in mind, the parties may well be inhibited in giving the mediator sensitive information during the mediation, and thereby lose much of the advantage of having a neutral third party assisting them in the negotiations
- The mediator's recommendation may be accepted by one, or neither, party and so the mediator's position of neutrality may be destroyed for the other, or both parties.
- The parties may abdicate responsibility for the problem and solution. Knowing that the mediator may be manipulated into giving a recommendation may change the focus from negotiating a deal to making the mediator take responsibility for the outcome.
- Having given a recommendation, it is likely that the mediator will not be seen as a neutral go-between in trying to broker a deal for a mediation that does not settle.
- Any recommendation may alienate a party's advisor if it goes against their opinion.

- It may be an easy option. Better to keep the parties focused on their solution rather than respond to a (ego-boosting?) request for a recommended outcome.

The mediator's focus should be on giving the parties the best opportunity to achieve a settlement. Their problem, their solution. The mediator's opinion should be irrelevant and a request for a recommendation, whether it be in the mediation agreement or not, should be resisted.

7.9 Mediator liability

There are very few areas where the mediator may be vulnerable to a negligence claim, and this is borne out by the very low professional indemnity insurance premiums. There are, as yet, no successful claims against a commercial mediator in the UK, but potential areas are

- giving an opinion
- writing the settlement
- breaking a confidence
- coercing a settlement

all of which are fundamental taboos for mediators. Mediators facilitate deals. They manage a process which is client-focused and based on confidentiality and which gives the parties the power to decide their destiny. So no mediator should be vulnerable to the above risks.

Chapter 7 in a nutshell

- Mediation offers certainty.
- The outcome is up to the parties, not the lawyers or the judge/arbitrator.
- Deals with dignity allow parties to leave with their heads held high.
- Requesting a mediator's recommendation changes the dynamic and focus of the mediation and the neutrality of the mediator.
- Have a draft settlement agreement available in preparation for the likely deal.
- Have specialists/experts on call if the agreement is likely to have tax or other implications.

- If there is any uncertainty about the settlement sticking, change it or ask for a cooling-off period.
- Don't be bullied, by either the other party or the mediator, into a deal.
- What can be learned from the dispute that will ensure that history does not repeat itself?

CHAPTER EIGHT
Roles in Mediation (Who Does What?)

This chapter looks at the roles of the main players in a mediation in more depth than elsewhere and pulls together much of what has been said in previous chapters. For the mediation process to be used to optimum benefit everyone needs to know their role and what is expected of them on the day. On the party side the key players are

- client (decision maker)
- legal advisor

and possibly

- counsel
- expert
- consultants
- support staff ('coal-face')

plus

- mediator
- assistant

or possibly

- co-mediator

8.1 Client

The decision maker's role is to negotiate the deal. To do this s/he must:

- Come with maximum authority. Ideally full authority to settle (that is, up to the full value of the claim/down to the offer currently on the table) but realistically authority is likely to be limited. If it is, then there must be a line of communication open to review the limit if this becomes necessary.

- Be clear and confident about the BATNA (best alternative to a nego-tiated agreement). Know what alternatives there are to settling at the mediation.
- Be prepared to tell his/her story and not leave it to the lawyers. Mediation gives a day in court that is much better than court itself.
- See the dispute as a joint problem that needs a joint solution. This means being prepared to hear the other side's story and cooperating in reaching a settlement to which they can also say 'Yes'.

It is often an advantage when the decision maker is someone who is not directly involved in the dispute itself. Mediations are settled through commercial negotiation, and being able to bring a detached and dispas-sionate approach to important decision-making is a real asset. This does rather conflict with the principles that mediation offers a day in court and that it provides an opportunity where emotions can be vented. However, it has to be said that the larger the firm, usually the less per-sonal the issues become for the decision maker, and so the day in court and emotional expression tend to be more relevant for the smaller and more personal disputes. That is not to say that people in larger firms do not get personally involved in disputes but rather that larger firms are usually able to offer decision makers who can be more removed from the centre of the dispute.

8.2   *Legal advisor*

Invariably parties in commercial mediations have legal advisors present. By that I mean solicitors – and often more than one. Most commonly it is a partner (to give authority to the proceedings) with an assistant (who has done the preparation work) and often a trainee (to get the experience). It is invariably of benefit to have solicitors at the mediation. Very few are negative or obstructive. Most encourage settlement and many are beginning to use mediation effectively. But it is a different forum to litigation and requires different skills. It should not be a matter of transferring the litigation approach into the mediation arena. The solicitor's role in a mediation should include the following:

- Supporter, not leader, particularly in the later stages of the mediation.
- Orchestrator – making it clear who does what and when.
- Analyser – helping parties to assess and reassess the risks and likely outcomes.

- Confident advisor – a well-prepared solicitor will be confident to support and advise his/her client on the legal framework of the dispute; on the strategy adopted by the client team; on the BATNA; on negotiating the best deal.
- Encourager, particularly when times are tough or the mediation appears to be doomed to fail. This is a tough process – if it were easy, they wouldn't need a mediator – but most mediations settle.
- Facilitator of his/her client's needs. In the end the deal is whatever the client, not the solicitor, decides, and only the decision maker knows why one deal is satisfactory and a different deal is not. Parties rarely tell all, even to their paid advisor, and many deals are done in mediation that leave the solicitor mystified as to why. The solicitor should therefore be prepared to stand back and allow his/her client to do what they think best, irrespective of legal rights.
- Positive influence by helping her/his party to look to the future rather than to wallow in the past. The past cannot be changed but mediation can form the future.
- Wise head. To be aware of a client's potential for getting into a face-losing situation and so steer them away before they get there. The solicitor knows his/her client better than the mediator does, and so is in the best position to anticipate such situations.
- Creative thinker – not least in using the idle time in a positive way, keeping the whole team engaged and active in seeking the best solution.
- Friend – not least to the mediator. Mediation can be a lonely business and the mediator is on the go throughout the mediation, so words of support and encouragement are always welcome.

This puts a big responsibility onto the litigation solicitor. It is not easy to adapt a lifetime of adversarial habit, of being at the front of the 'fight', into being cooperative and giving his/her client the freedom to do and say what they want. For a profession that is trained to be in control, to revel in the detail and to take pride in their negotiation skills, such a change can be a challenge indeed. But it is vital if the mediation process is to be used to best effect and the party is to get the best deal available.

8.3 Counsel

Chapter 4 mentions that more and more mediations have counsel present, but so often this is unnecessary and causes a significant extra expense that is rarely justified.[1] Almost always it is the result of one party bringing counsel and so the other parties feel the need to match and bring theirs. It is quite a brave solicitor who advises a client not to reciprocate when a party decides to bring counsel. The easy, and safer, option is to bring one as well. However, it is good to be brave, and experienced solicitors will recognise that mediation is not the best arena for counsel to use their skills. This is not a place for great, or even mediocre, oration, or even for argument on legal detail. Both will cloud the key reason for being there – to achieve a deal to which all parties can say 'Yes'.

But if counsel are to attend, to be truly effective they should be clear about their role. The instinct to lead, to speak, to control their client has to be resisted. Counsel are most effective in mediation if they are able to do the following:

- Advise. That is their natural role and it can be a great asset in mediation if they are able to maintain a position of independence amongst others who are inevitably partial and who have a keen interest in the outcome.
- Use their relationship with other counsel. Counsel almost always have a respectful relationship with each other and this can be an asset in mediation if negotiations get deadlocked or matters get heated. The mediator may well feel able to put counsel together to move the mediation towards settlement when other routes are blocked.
- Remain independent. This may be difficult because the counsel's role is primarily as advisor and their advice may well be different from that of the other party's counsel(s). It is most likely that counsel will feel the need to defend their opinion. However, if it is possible, a more detached and independent role can be a real asset to a party, both in reality-testing their position and by being a sounding-board for positions and potential offers.
- Avoid being their client's mouthpiece. This is the most challenging of all – the instinct is always to speak on behalf of their client. It is their natural role in a courtroom and a difficult one to shake off in a mediation. The mediator will want the parties to speak, and preferably to speak more that the legal team. I have yet to be in

---

[1] See report on the December 2007 Debate 'Barristers are giving mediation a bad name' on the MATA website www.mata.org.uk.

a mediation where counsel feel comfortable in taking such a back seat.

- Accept that mediations do not settle on legal argument. The legal framework does provide the backcloth to the dispute but mediation returns the dispute to a commercial stage where after a time the legal arguments cease to be key.

Having said all that, this is not a counsel-bashing exercise. As with everyone in a party's team, they need to be clear about their role and ensure that they use the opportunity to their client's best advantage.

8.4 Experts

Mediation does not provide the best forum for experts to shine. Usually, by the time they get to the mediation, each side has obtained an expert's report; the experts may well have met and agreed areas of common ground, and what is left are the areas on which they disagree. They are unlikely to change that just because they are in a mediation and their client must make a judgement on which of the experts is likely to sway the judge/arbitrator if the matter does not settle in mediation. Even if the experts have not met, and common ground may then be agreed within the mediation, it is rare indeed for experts to agree in total. The danger is that a party's expert becomes a barrier to settlement because they are so adamant that their version is the correct one. The party needs reasons to be flexible, not to become set in a position.

So if experts attend the mediation they need to do the following:

- Recognise that they may be idle all day. The mediator may well put the experts together but the outcome rarely affects the discussions, and the reason for putting them together may well be to give them something to keep them occupied.
- Avoid becoming a stumbling block to settlement. The expert needs to understand that his/her report is just one part – and usually a small part – of the overall picture and that their client will settle on commercial terms that may not reflect their opinion.
- Accept that mediation is not the forum for proving them right or wrong. So they should avoid being confrontational and recognise that cooperation is the best way of helping their client to achieve settlement.

Experts' reports are often important in construction disputes. They help to clarify the issues and identify the key areas of difference. But the important work is done well before the mediation and their best role in a mediation is to stay away, and perhaps be available on the telephone in case something does arise that needs their input. Alternatively, if they do attend, then it is worth ensuring that they have their say in the opening joint meeting and then suggest they leave, for it is almost certain that they will be asked to say little thereafter.

8.5 Consultants

There is a difference between experts and consultants. In this context I mean consultants as architects/project managers/engineers/valuers/quantity surveyors: people who have been involved in the project but who are not either the party or the expert. In some cases they may fall into the next category of supporters – people who were at the 'coal face' of the project and who have an interest in the dispute, and probably its outcome. Whether serious or not, the threat of professional negligence is invariably present. Decisions made during the project come under scrutiny and consultants find it hard not to become defensive. The fact that those decisions were probably made in the client's best interest at the time ceases to be relevant when the other parties in a dispute are seeking to lay blame on anyone other than themselves.

So, again, it is not always best for consultants to attend the mediation. They may well help in reinforcing their party's version of the truth, of recounting the history, but their use becomes limited in structuring a commercial settlement. After all, if the dispute is over delays and extension in time claims, an architect or project manager is unlikely, even for the purposes of the mediation, to grant a further extension in time. Similarly, a quantity surveyor who has been resisting claims or rates for work is unlikely, for the purposes of the mediation, to suddenly agree. So there is a danger that whilst, as with experts, their historic knowledge of a project may be useful, their usefulness becomes limited and potentially a barrier to settlement as the mediation progresses.

It is really important that the decision maker has the support of the team in any eventual settlement and sometimes it can be quite difficult for consultants to support a deal that concedes matters that they have resisted in the past.

So when consultants attend they should do the following:

- Recognise their own personal investment in their position. It is necessary to detach the person from the problem and try to achieve a position of independent professionalism.
- Similarly be aware of, and preferably discuss beforehand with their client, the influence that a blame culture brings to a dispute. Statements will always be guarded and defensive if there is a fear of a potential negligence claim.
- Support their client in any settlement that is reached in the mediation, even if it does go against their own advice.
- Be aware of 'face' issues that may influence their, and others', position.

The best approach to including consultants in a client team is to have a frank and open discussion about the issues beforehand and to agree a 'no blame' culture in which confidentiality prevails throughout the mediation.

8.6 Support staff

It was mentioned in an earlier chapter that teams should be 'lean and keen'. Parties often assemble teams as much for comfort as for their potential usefulness, and this leads to large and sometimes difficult teams – difficult to keep engaged in the process and difficult to keep united in purpose. However, it is probably important for those who were deeply involved in the project (site agent, project manager, foremen and so on) to tell their story and so have their say in the opening joint session. But they need to be briefed and to be aware of the fact that there is a likelihood that their strong feelings about how it really was will mean they may become an obstacle to a settlement. So if they attend the mediation, they should do the following:

- Be briefed about their role and clear about what they should say.
- Understand the potential downside of attending; that they will be deaf to other versions of the truth and be emotionally attached to what they see as justice.
- Support the decision maker in any settlement that is reached, even if it is contrary to their sense of justice. No decision maker wants to feel

members of the team will return to work saying that it was a poor deal.

- Accept that history cannot be changed and that commercial deals mean that everyone can return to generating wealth for the company rather than continue to spend valuable management time on pursuing/defending a claim.

Because mediation is a flexible process it would be possible for the 'coal-face' staff, consultants and experts to have their say on the first ('history') day and then have a much smaller team negotiate the deal on a second day, perhaps with a gap between the days to allow what has been said to be assimilated and positions to be redefined.

8.7   *Mediator*

Basically the mediator is there to give the parties the best chance of doing a deal. To do that the mediator should

- manage the process efficiently
- be clearly independent and even-handed
- build a relationship of trust and openness with everyone, but particularly with the parties
- maintain confidentiality so that parties will share sensitive information and know that it will not be used to their disadvantage
- encourage all parties to tell their story, acknowledging emotion and giving value to what they say
- re-open fractured lines of communication and sometimes help restore relationships that have been damaged by the dispute
- enable parties to tell (and listen to) their version of the truth
- help parties move from the past to the future
- sensitively test the reality of their positions and help them move from entrenchment to flexibility, from confrontation to cooperation
- help parties focus on the key issues and the commercial realities
- inject common sense into situations that have gone beyond the parties' control
- help parties avoid situations where they can potentially lose face
- ensure that settlements are realistic and will stick

In the end, though, it is the parties' problem and their solution. The mediator can give them the best opportunity for doing a deal but it is up to them how they use it. Some parties foul it up. Thankfully few mediators do.

8.8 Assistant

Assistant mediators are there primarily to gain experience of the real thing, no amount of simulation in their training being as good as actual cases. They are also there to do the following:

- Help the mediator. They are message takers and refreshment arrangers; they allow the mediator to bounce ideas off them and perhaps they even contribute ideas themselves. They time-keep when asked, and sometimes when not.
- Add a dimension. I like to have a female lawyer to add to my male non-lawyer so that we cover all angles.
- Be a friend to the mediator. Mediation is a solitary profession and sometimes the mediator has to soak up a lot of emotion and bad feeling. Parties often make the mediator's life unnecessarily difficult, so a friendly supporter can make a lot of difference and keep the mediator fresh, energetic and impartial.
- Perhaps chair a working group if the mediator wants to be with others.
- Spend an hour with the mediator after everyone has gone, to debrief the mediation. This not only helps the assistant bed down the learning points but also helps the mediator unwind from the mediation and so sleep soundly when s/he gets home!

The mediator chooses the assistant and, although they have the right to do so, it is rare for parties to object. Perhaps a potential conflict of interest if the assistant is known to a solicitor or party may lead to an objection. Otherwise, the assistant is bound by the same confidentiality provisions as the mediator, is not paid (although some mediators may pay the assistant's expenses so that they are not out of pocket) and is usually a real advantage to the process.

8.9 Co-mediator

It is worth mentioning co-mediation because it is particularly suited to multi-party disputes, and that means most construction disputes. There is a detailed paper on co-mediation on the MATA website[2] but the key points are as follows:

- Co-mediation not only makes more efficient use of the time but also brings another experienced mind to the process. This significantly reduces the amount of idle time, when the mediator is with another party in private meeting.
- Co-mediation needs a familiar pairing. The pair should be able to work in complete harmony and as a unit. Mediators are used to being in charge because they effectively work alone so if the pairing is ad hoc the danger is that there will be two competing egos. This will change the focus and potentially work against the parties' interests.
- There is an opportunity for having a pairing that suits particular cases: for example, lawyer/non-lawyer, generalist/specialist or respected name/experienced mediator.
- Communication between the mediators is key. They will be working separately for much of the time and need to know what is going on elsewhere. The arrangement that I have with my co-mediation partners is that, whatever we are doing, we meet every hour on the hour for five minutes to brief each other so that we can be sure we are moving in the same direction.
- Because two experienced mediators use limited time efficiently, co-mediation is cost-effective and the best option for multi-party disputes.

Other forms of mediation use co-mediation in a different way. Both matrimonial and community mediators work in pairs but are always together. In commercial mediation, the greatest benefit is when they work apart because it optimises the limited time available.

Chapter 8 in a nutshell

- Keep the mediation team lean and keen.
- Be clear about roles and about what is expected of each person.
- Ensure that the decision maker is supported by everyone, regardless of the eventual deal.
- The decision maker should come with full authority to settle.

---

[2] www.mata.org.uk (see 'Co-mediation' within 'Papers').

- All other members of the team attend to support the decision maker.
- The key to achieving a settlement is moving from history to influencing the future; from looking back to looking forward. Mediation gives everyone the opportunity to tell their story but, having told it, they then need to let it go.
- The mediator is there to give the parties the best opportunity to achieve a settlement, but in the end it is their settlement and how they use the opportunity is up to them. Some foul it up.
- Co-mediation should be seriously considered for multi-party cases. It is time-efficient and the extra cost is relatively small.

CHAPTER NINE
Avoiding Disputes in the Construction Industry

Conflict can be – should be – good. It is a catalyst. It can create dialogue, promote creative thinking, inspire people to sustainable solutions. Handled badly, though, it can create division, polarise positions, fracture relationships, harm business and send people off onto the inevitable spiral of dispute and litigation (or arbitration). And once on that spiral it is immensely difficult to get off.

9.1   The positive side of conflict

There is a critical point when conflicts can go one way or the other. I call it the decision point – the point where a person makes a decision to either fight or flee, or to take a breath and make an effort to enter into dialogue with the other person. Most people choose the former for any number of reasons. It might be fear, to avoid blame or admitting a mistake, to avoid nastiness or embarrassment. It might be a sense of injustice or the determination that the other person must not win. Whatever the reason, most people choose to fight or to walk away. And usually regret it later.

But what if, at that critical decision point, they choose to pause and remember that there is a better way, and bother to ask why the other person is seeing the situation so differently? It is the basis for starting a process of understanding, valuing and cooperating.

9.2   Understanding – valuing – cooperating

Dialogue can lead to understanding. Once a party understands where the other is coming from they may be able to accept that the other party's perspective, though different from their own, is genuine, so it should be valued. Once a person feels heard and valued, they are likely to cooperate in resolving the issue. So asking the question 'Why?' – Why are they taking a position that is different to yours? Why are they interpreting the same facts and events differently to you? Why is their truth different from yours? Why do they feel so strongly about it? – is the key to understanding.

Of course, there is no point asking the question if you are not prepared to listen to the answer. Properly listening, and demonstrating that you are doing so. Which should mean that you will begin to understand why they see the situation differently to you. Understanding does not mean that you have to change your view – but it might. Even if you do not agree with their interpretation you can value the fact that it is genuinely held. Showing that you respect their position is the next step to working out a solution that reflects their needs as well as yours. It leads to cooperation, working together to find a solution to which both can say 'Yes'.

It is worth repeating the three words: *understand* why – *value* their views – *cooperate* in finding a solution.

That is the positive side of conflict. It does not have to lead to war or capitulation. It can lead to a strengthening of relationships and pleasure at a jointly satisfactory outcome.

9.3   *Creating a culture that is positive*

Fight or flight is mentioned above as the usual response to conflict. For a positive and creative environment, one that nurtures and encourages innovation and risk-taking, the very opposite of the modern blame culture needs to be created. An environment where

- problems are not buried but are dealt with cooperatively
- mistakes are seen as learning points, not the cause for laying blame
- people are supported, not isolated
- there is a sense of common purpose and unity of vision

In such an environment people blossom and invariably enjoy their work. If they are in an environment of fear, they are demotivated and do the minimum, don't take risks and won't take responsibility. In a culture of blame there is no incentive to work together to resolve problems, only to protect your own position and ensure nothing detrimental goes on the next annual appraisal.

9.4 *Twelve rules and challenges*

One of the outcomes of mediating disputes is that the mediator often looks back at the dispute and thinks, 'If only they had said/not said that or done/not done that, it would never have become a dispute.' So it is inevitable that, over time, mediators will identify a 'rule book' for avoiding disputes. It may seem that we are intent on putting ourselves out of business but many mediators are keen to teach companies dispute avoidance techniques so that they may never need a mediator. Here are my twelve 'rules'.

9.4.1 Establish clear, simple and constant lines of communication

Poor communication is by far the major reason for problems escalating into disputes. The lines of communication need to be established from the start. They need to be as short as possible so that the messages do not get altered or misinterpreted. And who communicates with whom and how needs to be decided. Always remember that speaking face to face is the most efficient and effective form of communication. Research shows that, compared with face to face (where body language can supplement the words), speaking on the phone is only 38% efficient. With the spoken word, tone, pace, emphasis and silence all add to the meaning of the word itself. Unfortunately, we are in an age where most communication is by text or email – the most inefficient and impersonal method of all communication. The same research puts a 7% efficiency on the written word. How much more, then, it is prone to misunderstanding, misinterpretation and conflict-generation! So face to face wherever possible, and telephone if not. Then confirm by email or text if necessary.

It is often said that the meaning of what is said is in what is heard. In other words, it is not what you mean that is important, but what the listener hears. Really it should be taken one step further. It is the intention behind what is said that is important, and if the listener misunderstands then the speaker should be given the chance to rectify the situation. It is the intention behind the words that needs to be understood.

9.4.2 Establish clear roles, responsibilities, accountabilities and systems

Simplicity is everything, whether it be in systems or in language. The simpler it is, the less likely it is to be misunderstood or misused.

Decision-making should be devolved to the lowest level, both making decisions speedy and at the same time creating ownership and commitment throughout the lines of authority.

Site and other meetings are often lengthy and frustrating events and this is invariably due to poor and inefficient chairing. Meetings should be chaired by the most able rather than the most senior person. It doesn't have to be the architect partner or the project manager; much better a person who can manage the time well, and ensure everyone keeps on task but still has their say. Agendas should be realistic and pre-agreed to enable preparation, and support papers should be circulated well in advance so that they may be read and considered before the meeting.

Problems will occur. It is important to recognise this and agree an early warning system so that they can be resolved as soon as they arise. Problems ignored, or recognised too late, will damage both relationships and the desired outcome.

Record-keeping should be for everyone's benefit. The form of records and the distribution list should be agreed at the start of a project so that everyone has access to the relevant information and queries can be raised and resolved at the earliest stage. It serves no purpose to secrete information in the hope of gaining advantage – or in the fear that the other parties will know too much – at some later date. There should be a spirit of partnering and cooperation from the start.

9.4.3 Practise (and therefore model) openness/transparency

Following on from the last 'rule', openness means sharing information, including records, so that everyone is fully informed and able to contribute confidently. It means having no hidden agendas; being open and up-front, and so encouraging other people to lower their defences and to respond in a similar manner. Unfortunately it does also mean being prepared to be abused occasionally, but the rare occasions on which this happens should not cause the benefits of trust and cooperation to be lost. They far outweigh the potential disadvantages.

Many successful contracts have been run on an 'open book' principle – nothing to hide, no one taking advantage, no one losing out. What you see is what you get. What better model could there be for others to follow?

9.4.4 Build trust from the start; cooperate rather than confront

Good relationships mean good business. If the relationships are right, everyone will cooperate when problems occur. If they are wrong, everyone will become defensive and seek to apportion blame. Trust is built by allowing everyone to contribute, by being prepared to listen and value others' views, by being open and transparent. Trust is hard-earned, and difficult to recover when it is broken. So it needs to be treated as a precious commodity, to be nurtured and protected. It may be difficult and may fly in the face of the tradition of a generally combative industry, but parties should assume that each is to be trusted, and establish that as a rule from the beginning. Assume that everyone is there with good intent and that they are to be trusted. Assumption of trust, rather than adopting a cautious and suspicious approach, creates a totally different ethos and environment. If you assume that all are to be trusted, few people will risk abusing you. But this may challenge the conventional working of those who have long made the assumption that no one is to be trusted and that everyone is out to gain advantage, usually at your expense. Hang in there! Trust is good and should be the headline for all relationships on any project.

9.4.5 Acknowledge problems, don't bury them

Things do go wrong. Human beings do make mistakes, but not everyone feels able to admit to them. The worst thing is to hope that problems will go away if they are ignored. The likelihood is that they will get worse, and the fall-out will be greater the longer they remain unsorted. The best thing is to give problems early recognition; involve others, seek opinions and come to a decision that everyone understands and will therefore support (even if they do not agree). Even if they are right, unexplained decisions handed down from above are unlikely to be given enthusiastic support by those who have to put the decisions into effect. So if time does not permit consultation over a problem, giving a logic to the decision will generate understanding and support.

9.4.6 Treat mistakes as learning points, not blame-makers

Everyone makes mistakes. It is part of being human and it is how most of us learn. From a very early age we learn from doing, from testing boundaries, from stepping into the unknown, rather than from being told. And this means that sometimes we make the wrong decisions. This usually affects others, and in a work situation wrong decisions can be serious. But they happen, and when they do, time cannot be turned back and a different decision made. The best reaction is to ask, 'What have we learned from this, and how can we avoid it happening again?' The worst reaction is to follow the blame culture and seek a victim. A blame culture (something has gone wrong; someone must be blamed and they are going to pay) destroys innovation and the taking of risks. It stifles pioneers and creates a mundane, sterile and selfish society that is interested only in self-preservation. It is not the culture for a vibrant and growing business – or the culture for a successful project. Mistakes should be learning points and so help people, and businesses, grow.

Forgiveness is a word that is rare in business language. But mistakes, in whatever form they take, need to be accepted, learned from and then passed by. Forgiveness is often necessary in order for moving on to be possible. Forgiveness not only for the 'perpetrator' but also for the wronged. People need to learn how to forgive not only for the sake of others but also for their own sake. Otherwise they will be stuck in the conflict and meet new situations with the old conflict still in mind.

9.4.7   Get the 'headline' agreed

Headlines have been mentioned several times already. They set the touch-point of negotiations, of team-building and of successful medi-ations. They are the reference point for all that follows. So getting the common goals, the desired outcome, understood and agreed from the start not only ensures that everyone is heading in the right direction, but also means that there is an agreed reference point against which all other decisions can be measured. Yet so often it doesn't happen. People assume that everyone is there with the same vision, the same purpose, and it is only later, often too late, that it is realised that those visions and purposes are not the same. Asking 'Why are we here?', 'What is the pur-pose of this meeting/project/event?' is vital to its successful outcome.

9.4.8   Listen, and show that you have heard

Listening is often very difficult in a busy world. It is so easy to have your say and then switch off. Whether it be intentional or not, it is so easy to believe that yours is the only opinion worth listening to. Or, in the pressure of a busy world, to just not listen because other important matters are buzzing around.

But good relationships and trust are built upon being prepared to listen to what other people have to say. More than that, actually demonstrating that they have been heard by acknowledging or responding to or summarising what has been said. Summarising is a great way of demonstrating not only that the speaker has been heard but also that they are understood. In addition it gives the speaker an opportunity to adjust what has been said if the summary has not properly reflected what had been meant. When they feel heard, people feel valued and so wish to cooperate.

9.4.9 Establish what parties need, rather than what they claim

This is a classic mediation phrase – needs, not wants. It is not what they are claiming but rather what they need to be able to say 'Yes' to a deal. So it is in any interaction. People, especially English people, rarely say what they really mean. We sometimes say the opposite, sometimes hint or allude to something and expect the listener to be sensitive enough to pick up what lies under the words and what we really mean. So be prepared to enquire, to seek to understand why differences occur, why their truth is different to yours. Find out what their drivers – their pressures – are; they could be reputation, pressure from above (or at home), budgets, need to succeed or 'win'. Ask what they need to be able to resolve the issue. Ask, try to understand, then show that you value (though not necessarily agree with) their position and demonstrate willingness to cooperate in achieving a mutually satisfactory outcome – one that meets their needs as well as yours.

9.4.10 Involve a neutral early when disagreements are unresolved

This assumes that you hadn't read this book at the time, otherwise the sense of deal or project mediation would have been recognised and used to the full. But if no independent third party is already on the scene, if problems that cannot be resolved by those directly involved do occur and conflict begins to escalate towards a dispute, then find an independent third party to mediate a settlement. Mediation is an assisted negotiation and often an independent mind can bring fresh thoughts and ideas, and so unblock the deadlock. The worst thing that can happen is that disagreements or other problems are put to one side in the expectation that they will be resolved at the end of the project. Such situations not only badly affect relationships but also divert people's energy and attention from the fulfilment of the job, to the preparation for a battle. Project mediation provides an early warning system for problems. Without it, there needs to be an agreed protocol for recognising and dealing with problems as soon as they arise. Getting a neutral person to intervene at an early stage preserves relationships and enables problems to be resolved cooperatively before they cause lasting damage.

9.4.11 Re-evaluate agreements and headlines in the light of resolution

Assuming that all the above is adopted, there will still be problems and challenges for the project team. Early and effective resolution should not be the end of the matter. There should be a defined policy for building lessons from past conflicts and disputes into the working framework to ensure that similar situations or misunderstandings can be avoided in the future or identified and dealt with at a much earlier stage. As mentioned above, mistakes (and cooperative problem-solving) are learning points, and that learning needs to be bedded into future working practices. Agreements and goals need to be checked and re-evaluated in the light of experience, in a continuous process of growing and in the pursuit of excellence.

9.4.12 Re-commit to the relationship/contract

Even when conflict is handled well and positively, relationships can still be challenged or damaged. Conflicts and disputes can result in a loss of motivation and commitment to the work in hand. It is important to recognise this, and to devote time and resources to re-committing to and strengthening the relationships and aims of the project. It is a matter of everyone involved looking to the future rather than the past once an issue has been resolved. Resolution should be seen as success – the fact that people have worked together to solve a problem – and something that should reaffirm the relationships, ethos and headlines of the project.

This is not just theory. It is born out of (other people's, I am glad to say) painful experience. It is also common sense and so worth adopting into common practice until it becomes habit. Common sense is a good habit to get into.

9.5   *Partnering*

And partnering is common sense. It has been around a long time but is still seen as a 'new' process. The trouble is that it challenges tradition and so has had its sceptics. It is a process that still needs to be 'sold' to the industry.

Partnering is 'a long-term commitment between two or more organisations to achieve specific business objectives by maximising the effectiveness of each participant's resources based upon mutual objectives, an agreed method of problem resolution and an active search for continuous measurable improvements'.[1] In other words, working together for mutual gain. But the 'together' bit is important. It means that the partners adopt a common philosophy that applies to communication, behaviour, attitudes, values and practices, which may mean transforming organisational culture. That is quite a challenge in an industry that is fragmented and traditionally adversarial. It means

- creating an environment of mutual trust
- adopting common, rather than individual, goals
- sharing gain, and loss
- involving all members of the team in communication and decision-making
- eliminating blame culture
- having a common aim of continuous improvement
- agreeing a mechanism for early problem-solving
- open-book documentation

The results can be remarkable.

Case study

In Denmark two massive bridge-building projects were carried out with two significantly different outcomes. The Storebaelt bridge was built in 1998 and was late, over budget and had a poor safety record. However, two years later the Oresund bridge was completed nine months ahead of the contract completion date, on budget and with a remarkably low accident rate. The difference was that the

---

[1] NAO 2001.

Oresund bridge contract was a partnering project where the employer took an active lead throughout in setting the philosophy of trust and openness, where the dispute resolution process (in this case a dispute review board ... dammit!) was agreed and operated from the start, and where a 'Musketeer bonus' (one for all, all for one) was in place from the start. So problems were resolved as they arose and the focus throughout was able to be on the 'headline' – the early, profitable, safe and high-quality completion of the bridge.

Successful partnering does not happen by accident. Although it takes time for a cooperative culture to spread to all parts of the project team, this can be hastened by training[2] and people's preparedness for change. It is, after all, common sense and a very human process.

Chapter 9 in a nutshell

- Communication is everything.
- Good relationships mean good business.
- Conflict is good – it creates dialogue.
- Understand – value – cooperate.
- Blame culture stifles innovation and growth.
- Trust is everything.
- Mistakes are learning points.
- It is a person's intention that is important – we all say things that we regret.
- Forgiveness liberates both the forgiven and the forgiver.
- Common sense is a good habit to get into.

---

[2] For example the MATA Partnering Academy, www.mata.org.uk.

CHAPTER TEN
The Mediation Landscape

It is inevitable that a flexible process based on common sense will be developed into other areas. Inevitable too that it will be moved further up the construction process to the currency of the project (project mediation) and even to the negotiation of the project itself (deal mediation). The skills involved in helping parties reach a settlement – their settlement – are the same at the start as at the end. Establishing lines of communication, building trust, fostering cooperation and achieving the best deals are as valuable, indeed more valuable, at the birth of a project as in mediating disputes. And those same skills can be applied throughout the project to ensure that when problems occur they do not escalate into full-blooded disputes. The mediation process and skills can be used at the beginning, the middle and the end. How sensible! Yet how rarely does it happen.

10.1 Deal mediation

10.1.1 What is it?

Deal mediation is the introduction of an independent third party into the negotiation process, to help the negotiating parties achieve the best outcome. In 1999 Mike Hager wrote a paper[1] advocating deal mediation in international business and then developed the idea of a new profession, mistakenly as a new legal specialist (but he was then Director of the International Development Law Institute in Rome, so he can be forgiven – whilst some lawyers may have skills in facilitating business negotiations, it is rarely a natural or instinctive quality, whereas facilitators from a business background are likely to be more naturally suited to such a role). The principles of deal mediation are nevertheless

---

[1] Hager and Pritchard, 'Deal Mediation: How ADR Techniques Can Help Durable Agreements in the Global Markets', *Foreign Investment Law Journal* Vol. 14, Spring 1999.

sound. Any business brings together people from different cultures, and of different education, gender, age and profession. Already a situation ripe for misunderstanding. Add to that the dimensions of nationality, language and beliefs, together with different, or no, legal systems and varying attitudes to the law, and the barriers to achieving the best deals are numerous. The idea is that the deal mediator, someone with a clear mind and with no vested interest in the outcome, can turn stranger-negotiators into effective collaborators with the shared aim of achieving a deal that is best for both/all parties.

10.1.2 What does the deal mediator do?

As with dispute mediation, the deal mediator manages a process and develops an atmosphere of trust so that the parties will share sensitive information and know that it will not be used to their disadvantage. The framework is often identical to that of dispute mediation – a series of private and joint meetings, although they may well be spread over a much longer period and involve a changing group of people. The deal mediator will be able to achieve the following:

- Obtain agreement on the desired 'headline' (and write it up as the touch-point for future discussions). This should happen in any negotiation. It ensures that the negotiations start with a point of agreement and that the parties are focused on the same purpose. It may seem elementary, but many negotiations have failed because parties have different goals and are negotiating in ignorance of the other's needs and objectives.
- Be a cushion for differing negotiating styles, so reducing the chances of breakdown or deadlock arising from cultural or other misunderstanding.
- Foster effective communication and understanding, particularly by ensuring that assumptions are checked out and differences, particularly cultural differences, are explained, understood and valued.
- Focus on parties needs and interests. As in dispute mediation, the aim should be to meet the parties' needs so that everyone can say 'Yes' to the deal.
- Help parties distance themselves from emotional attachment to issues or principles. A common mediation statement is 'separating the

people from the problem'. Emotional attachment causes problems. Focusing on the desired outcome is the key.

- Reduce the risk of stalemate, particularly to ensure that parties do not place themselves in positions where they could lose face.
- Carry out frequent reality checks to ensure that expectations are managed and are realistic.
- Anticipate areas of future conflict and so prevent disputes in the future. Sometimes the deal mediator is retained after the negotiations are successfully completed to help resolve issues or disputes that may arise during the project.
- Devise options for mutual gain through fostering creative outcomes.
- Ensure there is nothing left on the table when the deal is done. The deal mediator is privileged with holding sensitive information from all parties and therefore knows better than anyone what the possibilities are for negotiating the best deal.
- Maximise the quality of the deal and its sustainability. Facilitated deals invariably lead to fair and wise outcomes.

Of all these benefits, the one that cannot be matched by even the most skilful and collaborative negotiator is that of the mediator being the trusted confidante of all parties in the negotiation. No one else has that privilege or advantage. This fact alone means that deals are better and more efficient, and maximise all the potential for structuring the best deal.

The downside of involving the independent third party, aside from the cost, is that the negotiations can be more protracted. The counter to that is that they are invariably more sustainable.

10.1.3 Difference between a deal mediator and dispute mediator

Although a facilitator, the deal mediator tends to be more pro-active in the pursuit of the outcome than a dispute mediator who manages the process and allows the parties to be in control. The challenge for the deal mediator is to stay impartial when championing creative solutions and ensuring that the parties are not pushed too hard into doing the deal. David Shapiro, in his article on deal mediation in 2003,[2] states that, notwithstanding this, neutrals engaged in deal mediation often venture

---

[2] *Journal of ADR, Mediation and Negotiation*, Vol. 1, Issue 4.

where parties cannot, and venture where mediators in ordinary dispute resolution dare not. How exciting!

10.2 *Project mediation*

In the late 1980s the Hong Kong airport project adopted an innovatory approach to dispute resolution. It was a massive project and predicted to have many disputes. So the parties arranged for a team of mediators to be trained. The mediators were then based on site throughout the project so that they were on hand to resolve any conflicts before they escalated into disputes. This experience gave mediation a high international profile but it was not until 2000 that project mediation, sometimes called contracted mediation, was introduced into the construction and IT industries in the UK. ResoLex[3] originally formalised the process, and more recently CEDR[4] launched their own scheme. Project mediation puts the mediation process into construction projects from the start. Its focus is more on dispute avoidance than on resolution but one of its great advantages is that it creates an atmosphere of cooperation and partnering. Wise business people know that when the relationships are right people work together to resolve problems. Spending time on the relationships is an investment in the success of a project. Project mediation does that through having a mediation pairing appointed from the very start. They then do the following:

- Attend (some) team meetings as the project develops.
- Run a partnering workshop for all the project team.
- Develop a dispute resolution plan so that everyone has an agreed procedure to follow when difficulties arise. This breeds confidence which in turn encourages the team members to resolve issues themselves.
- Attend site on a regular basis.
- Are available to chair difficult meetings. The project leader is not always the best person to chair meetings.
- Are called in to resolve problems that the team have not been able to sort out for themselves.
- Mediate unresolved disputes during the currency of the project or, at worst, at the end of the project.

---

[3] ResoLex www.resolex.com.
[4] Centre for Effective Dispute Resolution www.cedr.co.uk.

The key benefits of project mediation are as follows:

- It not only preserves but usually strengthens relationships. People do not like disputes but they do enjoy success, and having a harmonious working relationship leads to success.
- It prevents delays to the contract programme.
- It keeps the energy and focus of the team on the success of the project. Once disputes arise, the focus and energy tend to be diverted and the project suffers.
- The cost is agreed at the start. Any mediations are subject to a further fee, but they are rarely needed because the problems are resolved when they arise and so matters do not escalate into disputes.

It is therefore particularly suited to PPP/PFI-type contracts where parties are there for the long term and so good relationships, which can easily be fractured by conflicts, can be maintained despite the inevitable challenges that will arise. It fosters the collaborative contracting approach to projects envisaged by partnering, but the introduction of independent third parties into the project ensures that the undoubted advantages of partnering are not weakened by being left to be the responsibility of people whose focus is on other things. The mediators make it happen. It is a tragedy that this was not the preferred dispute resolution process for projects such as the 2012 Olympics. DRBs, the chosen dispute resolution route, are a far costlier, less flexible process.

The mediator pairing is usually bespoke: often a lawyer and a construction specialist or sector expert, and both of course trained mediators. The mediator skill of building rapport means that an atmosphere of trust is built quickly and team members cooperate from the outset. The mediators become members of the team and are seen as allies rather than as a threat. The project is bound to benefit.

So why isn't it used more often? I think the early resistance was due to the eternal optimism of the UK construction industry. 'There will be no problems on this project so why should we pay for the service?' It was seen to be a form of insurance, an additional cost on already tight budgets. But its reputation is growing, and the more it is used, the greater that reputation will become.

10.3 Dispute mediation

If you don't know what this is, then go back and read the earlier chapters!

10.4 *Facilitation*

Facilitation is the introduction of a third party (not necessarily a to-tally independent one) to ease effective communication. Some media-tors have developed a reputation for using their mediation skills outside the dispute resolution arena, for example in chairing difficult meetings or even being the independent catalyst for enabling participants to con-tribute to important discussions in an unthreatening environment. Fa-cilitation involves significantly more preparation than most commercial mediations. The mediator has influence over who attends, the prepara-tion of the agenda, and often over the venue and the framework of the day. There is a big dependency on the personality of the mediator – one who creates trust very quickly and who is sensitive to the moods and needs of the participants. But they are core attributes of an effective mediator anyway. The aim is to ensure that

- the right people attend
- everyone contributes
- the outcome is sustainable

10.4.1 Preparing for the facilitation

As with dispute mediation, early telephone contact is vital in the prepa-ration for facilitation. But in contrast to dispute mediation, the contact is with the parties direct. This puts great importance on the mediator's telephone manner because without the face-to-face contact, there can be no reading of facial expressions and body language. So tone of voice and pace and matching words are all very significant. If there are to be a lot of people in the meeting (say, 20 or more) individual contact is not likely to be possible with everyone. The important thing then is to es-tablish who are the likely blockers to the discussions and to spend time in speaking to them. It is important that the dissenting voice is given time but also that such a voice has a positive, rather than destructive, effect on the dialogue.

Having touched base with all, or most, of the participants, the medi-ator may well be the one to issue an invitation for them to attend the meeting, which may be held off site. If that is possible then, as with other forms of mediation, a venue that has natural light and external space where people can walk and work is very important.

10.4.2 Agenda

The agenda for a facilitation is usually set before the day(s), but often it can be far better to do this at the start of the meeting. This means that everyone has an input and so is not only able to include what is important to them, but also buys into the process.

The first action must be to agree the headline – what is the desired outcome of the facilitation? Agreement on that will dictate the agenda. As with negotiation, it is the touch-point for everything that follows. It is also interesting that an agenda set on the day by the participants very often deals with what would otherwise have been Any Other Business in a pre-set agenda! What is less important to some may be the key issue to others.

10.4.3 Structure of the day

As with mediations, the day will open with everyone together. It is not usual to have private sessions beforehand because this is rarely about separate and distinct parties, but about a number of people all of whom have opinions (often all different) about the issues being discussed. Part of the objective of this first meeting will be to agree a strategy for the day(s) and a timetable. There are likely to be working groups during the day, but it is unusual for the mediator to have separate private meetings. These may well happen if there is a gap between main meetings, but they are rare on the day itself.

10.4.4 Open space

There is a form of facilitation called 'open space' that starts with a blank canvas. This is a much freer form of facilitation where conversation groups take place, on different topics, and people move from one to another depending on what interests them. The outcomes of the conversations are fed back to a central position and the results typed up and shared in a group plenary session. Although at times this process appears to be chaotic, it is held together by the mediators/facilitators who give the event structure and purpose and instil some calmness and direction. Users of this form of facilitation maintain that much richer outcomes result because people engage in the issues that really interest them and feel able to speak more freely in what is generally a more relaxed and informal forum.

10.4.5   Role of mediator as facilitator

The skills and much of the process of mediation are the same in facilitation but the emphasis is different. The mediator brings structure to the process and ensures that everyone has a voice. S/he will provide a safe environment to enable people to speak without fear of reprisal or discrimination as the object is to achieve the headline goal. Everyone is there for the same reason – to achieve the headline agreed at the start. The mediator ensures that the time is used efficiently and that everyone is focused on the task. S/he will challenge and reality-test just as in mediation. The whole purpose is to give those present the best opportunity of achieving the agreed desired outcome.

10.5   *Consensus-building*

Although this is only a brief mention, it is a natural extension of mediation to include consensus-building. Whilst most construction disputes occur during the progress of the work, and so consensus-building is not necessary, turn-key packages, PPP/PFI projects, change of use, brownfield and other developments often have an element of public consultation and debate. When this happens it can be a nightmare trying to manage the various and conflicting opinions and agendas. The process needs careful thought and handling so that the end result does not create a bigger problem than existed before the consultation – there have been many instances in which a public consultation has caused individual voices to combine into powerful protest groups because the consultation process has been mishandled.

Although most public consultations are bespoke, there are some basic rules very similar to those in facilitation that apply whatever the process:

- The headline must be agreed. What is the desired outcome?
- The constituency must be identified. Who should be involved? It may well be far beyond those immediately involved.
- Everyone should feel that they have been heard. This is the only way to remove blockages and generate an atmosphere of cooperation.
- Groupings should be formed that broadly represent the various positions. This is the only way to manage a large number of people with diverse interests.
- A process should then be agreed to achieve the desired outcome. This could be mediation or a series of facilitated discussions or even an agreed action plan.

Dialogue by Design[5] has devised an on-line consensus-building programme to supplement its paper-based and face-to-face facilitation processes. Much of the above is suited to electronic gathering of information, and Dialogue by Design have a long and successful track record of consensus-building on large public projects such as the Newbury bypass, the transport of hazardous material and the disposal of redundant oil rigs.

Effective consensus-building relies on many of the mediation skills covered above, particularly active listening, effective questioning, being non-judgemental, accepting people where they are and moving them from the past to the future.

10.6   Bespoke mediation processes

10.6.1   Construction Conciliation Group (CCG)

The CCG was formed to meet a need for domestic building disputes to be resolved in a sensible way. The right for adjudication did not exist and so most cases of relatively small value went to litigation. Often the legal costs exceeded the value of the claim, a situation that no one, lawyers included, enjoyed. The CCG devised a package which offered

- fixed fees
- fixed time
- certainty on outcome

Under the scheme the parties endeavour, with the assistance of a mediator/conciliator, to reach an amicable settlement within an agreed fixed period (say, five hours). If no settlement is achieved, the mediator/conciliator makes a binding decision. This decision remains effective unless and until the matter is determined by arbitration or litigation. The CCG panel of mediator/conciliators and their details are published on the CCG website[6] and users choose from the panel and contact the mediator/conciliator direct. If parties cannot agree then the CCG appoints.

This scheme, though devised specifically for domestic building disputes, can be used for disputes in any sector.

---

[5] www.dialoguebydesign.net.
[6] www.ccgroup.org.uk.

10.6.2 RICS Neighbour Dispute Service

In 2007 the Royal Institution of Chartered Surveyors[7] launched its Neighbour Dispute Referral Service. Following publicity on several neighbour disputes, particularly over boundaries, that sometimes involved assault, imprisonment and even death, the RICS devised a three-stage scheme to prevent matters escalating and to provide an alternative means of resolution other than through the courts. The stages are as follows:

- Evaluation. The surveyor provides the parties with an early assessment of the dispute. This involves a visit to the property, meeting with the parties, reading the deeds and writing a report. The report may include a recommended strategy for a resolution.
- Conciliation. The surveyor identifies potential areas of compromise and mediates with the parties. The term 'conciliation' is used as the surveyor may well be more assertive in advocating a solution than a facilitating mediator would be. If a solution is reached the surveyor drafts heads of agreement for the parties to sign. If applicable, the surveyor will mark out the agreed boundary.
- Expert witness report. Should the matter not settle in mediation, the surveyor will, at the request of either party, produce an expert witness report. This will be a full report to be produced before the courts. The report may refer to the parties' conduct during the previous stages.

Each stage has a fixed fee, paid equally between the parties. Although aimed mainly at boundary disputes, the scheme is also suited to disputes concerning

- rights of way
- party walls
- right to light
- trees
- noise
- parking and street issues
- riparian rights (river bank)

These may well encroach on areas of dispute normally mediated by community mediators, although they are obviously suited to surveyors who specialise in these areas in their normal course of work.

---

[7] www.rics.org

10.7 *Tiered resolution*

Some (wise) lawyers build a tiered dispute resolution process into contracts:

- Negotiation is the cheapest, most flexible and usually the most efficient way to sort problems out, and this would be specified in the contract. It may even provide for a tiered negotiation process, with the ultimate tier being a director of each firm. Making the parties talk to each other in an effort to negotiate a settlement by building this into the contract is more effective than leaving it to the parties to decide at the time.
- Assisted negotiation (mediation) would be the next tier. If straight negotiations fail then introducing an independent third party is the next best, and next most cost-effective, route.
- If mediation fails (and it rarely does) then adjudication would be the third level. Not so flexible, with an imposed solution usually based upon legal argument, adjudication takes the power for the settlement away from the parties.
- Finally, arbitration or the courts. Totally inflexible and costly, with an imposed solution based upon the arbitrator's/judge's interpretation of the law. There is a (usually not so happy) winner and a (very unhappy) loser.

The real value of a tiered resolution process is that it is there in the contract, and parties have to follow it. It makes them face the problem early, and that gives them a much better chance of resolving the issues before they escalate. Knowing it is there often means that they do not have to consider the second or later tiers, because they speak and settle.

Chapter 10 in a nutshell

- Mediation is an assisted negotiation. It has many more uses than just dispute resolution.
- The skills of the mediator are transferable and can add value to virtually any business situation.
- Adding a clear and independent mind can ensure that the best solutions are reached.
- The mediation process and skills are completely flexible.

CHAPTER ELEVEN
Conclusion – How to Win at Mediation

This concluding chapter is intended to summarise the key points of the previous chapters, highlighting how to use the mediation process – and the mediator – to best effect. Everyone can win in mediation, if they use it well.

11.1   Prepare well

Good preparation means entering the mediation with confidence – confidence in the process and confidence in the strategy to use it to best advantage. In addition to being fully conversant with the documents, good preparation includes carefully choosing the team to attend the mediation (decision maker, lawyer advisor and few others) and each being fully briefed on roles and who says what and when. Good preparation includes carrying out a detailed risk analysis including the alternatives to, and cost of, not settling in the mediation. It also means preparing to adopt, if only for a day, an approach that is different from the legal rights-based negotiation.

11.2   Chose the right mediator

There are a lot of mediators in the commercial market, some of whom have had minimal training and who are 'accredited' by right rather than ability. It is worth investing time in researching proposed names to ensure that the mediator chosen is not selected just because someone knows them, or has used them before. Experienced lawyers usually group their mediator database according to their client's perceived need – head-banger, good people skills, lawyer/non-lawyer, evaluative/facilitative, sector expertise and so on. The most suitable mediator may be all of these, or none. A mediator with good people skills does not have to be a softie. The best mediator would generally be someone with experience – and therefore someone who had an established reputation – who is a good negotiator and who can see the 'big picture', and who keeps the responsibility for the problem and solution with the parties. In other words, me!

11.3   Get the best out of the opening joint session

More and more lawyers are suggesting that the opening joint session be dropped and that the mediator goes straight to the private meetings, usually on the basis that 'we know their case, there is no need to repeat it all', or 'my clients feel very strongly, so it would be better for them not to be face to face with the other side'. Both excuses are so wrong. The opening joint session is the very place for parties to tell their story – not in cold legal argument but in terms of how they see it, how they have been affected, why they feel so strongly. It is likely that they will show emotion in doing so. And why not?

The opening session is an information-sharing exercise and so it is worth preparing questions beforehand and not letting the session end before each party can say, 'I know exactly what their case is about and they know exactly what mine is about.' Once parties are in their private room it is very difficult to ask questions and get useful responses.

Finally, be prepared to listen. It is so easy to focus on your own version of the truth and assume that any other version is wrong. It may be wrong, but it is worth listening and showing that you are hearing what is being said, because it might be a better truth than your own!

11.4   Cooperate

Mediation provides a forum different from any other. The best way to use it is to see the dispute as a joint problem and to endeavour to solve it by reaching a joint solution. The best deals come from cooperation because it enables each party's needs to be known and met and for all possible ingredients to a settlement to be on the table and used up in the eventual deal. This approach also enables relationships to be restored and the seemingly inevitable demonising of the other party to evaporate.

Mediation is the forum for wise negotiators. There is no need to – indeed every reason not to – be positional and 'tough'. Wise negotiators take the opportunity to explore all possibilities, including the non-financial, and establish a zone of likely settlement before tabling an offer. Mediation is a forum for returning common sense to business negotiations, and common sense is one of the wise negotiator's assumed qualities.

11.5   Have a drink!

When the deal is done – and it usually is – celebrate! Preferably with the other side, because they should be pleased with the outcome too.

11.6   And remember ...

- People see the same events/facts through different eyes. It doesn't have to mean that they are any more right or wrong – just different.
- As a result, people believe their version of the (same) truth.
- Mediation allows parties to tell their story (truly to have their day in court).
- The mediator is there to give the parties the best shot at doing a deal.

Appendices

Summary of Relevant Law

What follows is a schedule of the cases that are relevant to commercial mediation, prepared by Chris Cox, solicitor and mediator.

Table 1
Alternative dispute resolution pronouncements and recent mediation case law summaries

Date of Judgment	Case Name	Court/Judge(s)
1889	*Walker v. Wilsher* 23 QBD 335	Queen's Bench Division Bowen LJ
Remarks: Letters or conversations written or declared to be 'without prejudice' cannot be taken into consideration in determining whether there is good cause for depriving a successful litigant of costs. 'In my view it would be a bad thing and lead to serious consequences if the Courts allowed the action of litigants, or letters written to them "without prejudice", to be given in evidence against them as material for depriving them of costs. It is most important that the door should not be shut against compromises, as would certainly be the case if letters written "without prejudice" and suggesting methods of compromise were liable to be read when a question of costs arose.'		
8 March 2000	*Paul Thomas Construction Limited v. Hyland & Anor*	TCC Judge Wilcox
Remarks: Unreasonable conduct and failure to follow pre-action protocol leads to indemnity costs.		
23 March 2001		Lord Chancellor
Remarks: Lord Chancellor issues a formal written pledge that: 'Government departments and agencies make these commitments on the resolution of disputes involving them. Alternative dispute resolution will be considered and used in all suitable cases wherever the other party accepts it.'		

(cont.)

Table 1 *(cont.)*		
Date of Judgment	**Case Name**	**Court/Judge(s)**
8 January 2002	*Frank Cowl* v. *Plymouth City Council*	Court of Appeal Lord Woolf

Remarks:

A refusal to consider ADR (independent complaints procedure) is unfortunate.

February 2002		Civil Procedure Commercial Courts Guide. Draft ADR Order. To be found in Volume 2 of the White Book 2A – 163

Remarks:

Aside from compulsory mediation orders issued in the London Civil Justice Centre, the order below forms the strongest form of encouragement short of court compulsion.

DRAFT ORDER

'1. On or before [*] the parties shall exchange lists of 3 neutral individuals who are available to conduct ADR procedures in this case prior to [*]. Each party may [in addition] [in the alternative] provide a list identifying the constitution of one or more panels of neutral individuals who are available to conduct ADR procedures in this case prior to [*].

2. On or before [*] the parties shall in good faith endeavour to agree a neutral individual or panel from the lists so exchanged and provided.

3. Failing such agreement by [*] the Case Management Conference will be restored to enable the Court to facilitate agreement on a neutral individual or panel.

4. The parties shall take such serious steps as they may be advised to resolve their disputes by ADR procedures before the neutral individual or panel so chosen by no later than [*].

5. If the case is not finally settled, the parties shall inform the Court by letter prior to [disclosure of documents/exchange of witness statements/exchange of experts' reports] what steps towards ADR have been taken and (without prejudice to matters of privilege) why such steps have failed. If the parties have failed to initiate ADR procedures the Case Management Conference is to be restored for further consideration of the case.

6. [Costs]. Note: The term 'ADR procedures' is deliberately used in the draft ADR order. This is in order to emphasise that (save where otherwise provided) the parties are free to use the ADR procedure that they regard as most suitable, be it mediation, early neutral evaluation, non-binding arbitration etc.'

Date of Judgment	Case Name	Court/Judge(s)
22 February 2002	*Susan Dunnett* v. *Railtrack plc*	Court of Appeal Lord Brooke LJ

Remarks:

An unreasonable refusal to mediate will lead to uncomfortable cost consequences. The Courts show their willingness to give significant weight to the willingness or otherwise of the parties to attempt alternative dispute resolution pursuant to Civil Procedure Rules 1998, Rule 44.3.

9 May 2002	*Hurst* v. *Leeming*	Chancery Division Lightman J

Remarks:

If there is no realistic prospect of settling the case in mediation because of the obsessive character and attitude of the other party, then a refusal to mediate will be justified. Burden of proof reversed in Halsey. The case sets out the circumstances when it may be justified to refuse to mediate. The case is reviewed in Halsey (see below).

July 2002		Office of the Deputy Prime Minister

Remarks:

Office of the Deputy Prime Minister draft circular on Best Value and Performance Improvement states:

'It is in everyone's interest to work at avoiding contractual disputes in the first place and this is mirrored in the emphasis above on improving relationships between the client and contractor through team work and partnering. However, when disputes do occur it is important to have a fast, efficient and cost effective dispute resolution procedure. Local authorities should seek wherever appropriate, to provide clauses in their contracts on the use of alternatives to litigation (commonly termed Alternative Dispute Resolution) which can achieve this.'

1 November 2002	*Société Internationale de Télécommunications Aéronautiques SC* v. *Wyatt Co. (UK) Limited and others* (v. *Maxwell Batley (A Firm)* part 20 defendant)	Chancery Division, Park J

Remarks:

A party that was successful in the litigation should not be deprived of its costs because it reasonably refused to mediate.

(cont.)

Table 1 (*cont.*)

Date of Judgment	Case Name	Court/Judge(s)
9 June 2003	*Dearling* v. *Foregate Developments (Chester) Limited*	Court of Appeal, Civil Division

Remarks:

Held that the court had a power to make an award of costs even where parties have settled without a trial. The case therefore offers no encouragement to litigants to defend hopeless cases up to the door of the court in the belief of no order being made as to costs by the court if a commercial settlement is reached at that stage.

| 1 July 2003 | *Corenso (UK) Limited* v. *Burnden Group Limited* | High Court, Queen's Bench Division HHJ Reid QC |

Remarks:

Both parties found by the Court to have shown a genuine and constructive willingness to resolve the issue between them and therefore neither party should be penalised in costs for not having gone along with the particular form of ADR proposed by the other.

| 1 October 2003 | *Thakrar* v. *Thakrar* | Court of Appeal |

Remarks:

Tomlin Order (agreed at a mediation) refused at first instance. Upheld on appeal.

| 11 October 2003 | *Cable & Wireless plc* v. *IBM United Kingdom Limited* | QBD, Commercial Court Coleman J |

Remarks:

This case is a variation on the principle that an agreement to agree cannot be enforced. It decided that if you have agreed to a mechanism by which to attempt to resolve disputes, you can be compelled to follow that process. A dispute escalation clause was valid and enforceable.

| 5 December 2003 | *Shirayama Shokusan Company Ltd & Ors* v. *Danovo Ltd* | Chancery Division Blackburne J |

Remarks:

The court has jurisdiction to order an unwilling party to mediate its dispute. No longer good law (see *Halsey*).

Date of Judgment	Case Name	Court/Judge(s)
11 May 2004	*Halsey* v. *Milton Keynes General NHS Trust; Steel* v. *(1) Joy (2) Halliday*	Court of Appeal Ward LJ, Laws LJ, Dyson LJ

Remarks:

The burden is on the unsuccessful party to show why the general rule on costs should be departed from. The fundamental principle was that the normal costs rule applies unless the successful party had acted unreasonably in refusing to agree to mediation.

Date of Judgment	Case Name	Court/Judge(s)
27 May 2004	*Couwenbergh* v. *Valkova*	Court of Appeal, LJ Ward, LJ Waller, LJ Hale

Remarks:

This case contradicts previous rulings that fraud cases were not suitable for mediation.

'52 ... The parties had it, and still have it, in their power to alter the destiny of this appeal and this sad case. We urged them, and continue to urge them, to do so through mediation. It is a case crying out for alternative dispute resolution.

54. When costs do finally have to be allocated, we hope these observations will be borne in mind when the court comes to apply the guidelines in *Halsey* v. *Milton Keynes General N.H.S. Trust* EWCA [2004] Civ. 576 on how to deal with failures to mediate despite the encouragement to do so.'

Date of Judgment	Case Name	Court/Judge(s)
15 June 2004	*Reed Executive plc* v. *Reed Business Information Limited*	Court of Appeal on Appeal from Chancery Division Auld LJ, Rix LJ, Jacob LJ

Remarks:

The rule in *Walker* v. *Wilshire* (1889; see above) remains good law and the court cannot order disclosure of 'without prejudice' negotiations against the wishes of one of those parties to those negotiations. This means that when it comes to deciding the question of costs the court cannot decide whether one side or the other was unreasonable in refusing mediation. But the court went on to say that was not disastrous or damaging from the point of view of encouraging ADR.

Jacob LJ explains:

'Far from it. Everyone knows the Calderbank rules. It is open to either side to make open or Calderbank offers of ADR.

The reasonableness or otherwise of going to ADR may be fairly and squarely debated between the parties and, under the Calderbank procedure, made available to the Court but only when it comes to consider costs.'

(cont.)

Table 1 *(cont.)*		
Date of Judgment	**Case Name**	**Court/Judge(s)**
8 March 2005	*Bowman* v. *Fels*	Court of Appeal, Brooke LJ, Manse LJ, Manse LJ, Dyson LJ

Remarks:

This is an authoritative guidance as to the position of litigators under the Proceeds of Crime Act (POCA) 2002. This is a welcome clarification confirming that *the disposal of proceeding by consensual process in the context of civil litigation is also outside the scope of s.328 and just an ordinary feature of the conduct of civil litigation.*

Given the Court of Appeal's concern to encourage settlement, it is presumed that s.328 of POCA does not apply to any lawyer receiving information at the early stages of a dispute. But do take note that consensual arrangements independent of litigation could be *'an arrangement'* under the section which carries criminal sanctions.

8 April 2005	*Burchell* v. *Bullard*	Court of Appeal Lord Justice Ward, Lord Justice Rix

Remarks:

The alleged unreasonable conduct, a refusal to mediate, took place before May 2001, and was based on the advice of a surveyor. It was before the law was settled on the subject of cost penalties for failing unreasonably to mediate in *Dunnett* v. *Railtrack* in 2002 so the Court of Appeal (Lord Justice Ward) did not in this case penalise the refusal but instead issued the following warning to those who might unreasonably refuse offers in the future.

'The profession can no longer with impunity shrug aside reasonable requests to mediate. The parties cannot ignore a proper request to mediate simply because it was made before the claim was issued. With court fees escalating it may be folly to do so. I draw attention, moreover, to paragraph 5.4 of the pre-action protocol for Construction and Engineering Disputes – which I doubt was at the forefront of the parties' minds – which expressly requires the parties to consider at a pre-action meeting whether some form of alternative dispute resolution procedure would be more suitable than litigation. These defendants have escaped the imposition of a costs sanction in this case but defendants in a like position in the future can expect little sympathy if they blithely battle on regardless of the alternatives.'

Date of Judgment	Case Name	Court/Judge(s)
17 November 2006	*P4 Ltd v. Unite Integrated Solutions plc* Part 1 [2007]BLR pp1-10, February 2007	Queen's Bench Division (TCC) Mr Justice Ramsey

Remarks:

Helpful decision of the TCC on the effect on costs of a refusal to mediate and a failure to provide information at the pre-action stage. Ramsey J held that the defendant's failure disentitled it from costs to which it would otherwise have been entitled.

From paragraph 41 of the judgment: 'Experience of mediation has shown that the vast majority of cases are capable of settlement and are, in fact, settled in this way. In my judgment, that has to be the starting point.'

11 December 2006	*Finster v. Arriva and Booth.* SCCO ref CCD06040044	Supreme Court Costs Office Deputy Master Victoria Williams

Remarks:

Settlements reached where costs are not quantified but left that they are to be assessed by the court costs office if not agreed will always carry with them the very high risk that the paying party will use arguments about the case to argue as to the reasonableness and proportionality of the costs in order to reduce the sum to be paid to the receiving party.

24 May 2007	*Framlington Group Ltd, Axa Framlington Group Ltd v. Ian Barnetson* [2007] EWCA Civ502	Court of Appeal. Auld LJ.

Remarks:

'The claim to [without prejudice] privilege cannot turn on purely temporal considerations.' Auld LJ stated that the critical feature is the subject matter of the dispute and that one must ask whether in the course of negotiations the parties contemplated, or might reasonably have contemplated, litigation if they could not agree. Auld LJ therefore allowed the appeal holding that the parties were obviously in dispute and were both clearly conscious of the potential for litigation if they could not resolve the dispute without it and ordered that references to without prejudice discussions in a witness statement be removed.

Risk Analysis Checklist

This checklist was prepared by Roger Tabakin, accountant and mediator. It has been prepared for parties to disputes and takes the assessment of risk beyond the normal legal issues. Roger says: 'The skills required in preparing the risk analysis are likely to involve the significant use of subjective views, which are likely to alter as the information exchanges occurring in the mediation process arise. New information received [during the mediation] often leads to a change in perceptions.'

Client information sheet – mediation risk analysis

Normal transaction costs of filing a formal court claim and proceeding to the door of the court	Applicable to me Y/N	Estimated £ value best to worst	Applicable to other disputants Y/N	Estimated £ value best to worst
1. My legal fees				
2. My expert witness's fees				
3. Possible costs order against me				
4. Weeks of absenteeism from work preparing for court				
5. Weeks of lost employee time preparing for court				
6. Lost business opportunities for … months/years resulting from lack of concentration and focus at work				
7. Negative publicity in press or business circles				
8. Loss of control over my life to professionals				

9. Post-litigation recriminations against courts, experts and lawyers				
10. Loss of value by court ordered determination/valuation etc.				
11. Interest lost on money received later rather than sooner				
12. Months of personal stress and uncertainty				
13. Months of stress on family members				
14. Months of stress on my work associates				
15. Inability to 'get on with life' for … months/years				
16. Embarrassment and loss of goodwill if relatives/friends/ business associates are subpoenaed				
17. Cost and repeat of all previous factors if there is an appeal				
18. Lost future goodwill with and 'revenge' by opponents				
19. Estimated total of transaction costs (best to worst)				

NB: These are only rough estimates. All these figures will fluctuate up or down as the conflict develops and as more factors emerge.

Decision Tree

Dr John Clark, consultant, mediator and proponent of decision trees, writes:

'You need to choose between going to court and accepting a settlement offer. Going to court has risks attached. Using decision trees you can better understand the value of the court option relative to settlement offers. Understanding how the other side views their choices can help you in your bargaining strategy. Decision trees can be complex and need to be built with a clear understanding of the mechanics, where the numbers come from and their potential shortcomings. However, doing decision tree scenarios in advance is a vital tool to understanding your bargaining position and that of the other side.'

Decision tree example 1

- Your client has been defrauded of a large sum of money. You are pursuing the bank which paid the money out.
- You and the team have spent a lot of time assessing the pros and cons of your case and going through spreadsheets of numbers, accounts and money trails. The time has come to decide whether or not to accept the settlement on offer.
- The other side has offered to settle for £8m.
- You feel if you go to court you have an 80% chance of winning.
- If you win, there are a range of payments the court might make:
  - You estimate the best outcome would be £20m (after irrecoverable costs), but you think the chances of this are only 20%.
  - Most likely is £16m, with a 60% probability.
  - You think the chances of a low net payment of £12m are about 20%.
- If you lose you estimate you'll have to pay about £2m in costs.
- **Should you settle at £8m? If not, what should you settle for?**

Solution 1:

The award is £16m (£20m × 20% + £16m × 60% + £12m + 20%), which is then factored by risk at court (£16m × 80% − £2m × 20%) = £12.4m.

£12.4m is significantly greater than £8m; therefore go to court.

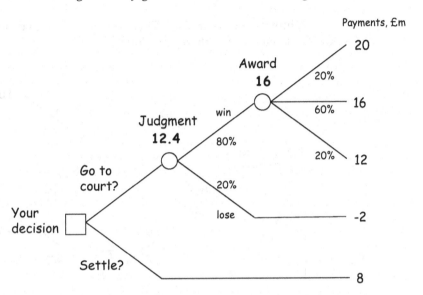

Decision tree example 2

- You have reappraised the situation. You have determined that winning in court depends on two decisions by the judge:
  - whether he allows the evidence your private investigators gathered as admissible;
  - whether he finds the bank dishonest.
- If he allows the evidence, then there is an 80% chance he will also find the bank dishonest and you win.
- If not, then it is only 50–50 he will find the bank dishonest.
- You currently believe there is only a 20% chance the judge will accept the evidence.
- The payoffs are the same as before.
- **Should you accept the offer?**

Solution 2:

Additional calculation is factored in. £12.4m from Solution 1 is adjusted by new risk analysis (£12.4m × 20% = £2.48m), then added to the additional calculation (£16m × 50% − £2m × 20% = £7m). £7m × 80% + £12.4m × 20% = £8.1m.

£8.1m slightly greater than £8m. Therefore, with risk aversion, accept certain settlement rather than go to court.

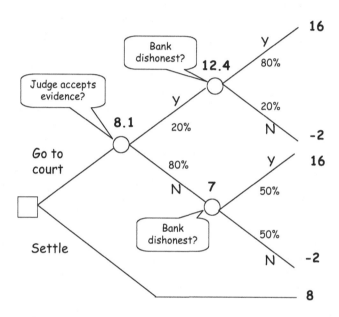

Pre-mediation Checklist

This is one that I use when contacting parties' lawyers before the mediation. It is a useful guide for parties in preparing for the mediation.

Pre-mediation contact checklist

NOTE: *Assumes confirmatory email already sent enclosing draft mediation agreement and covering date, starting time, document exchange timetable, venue and fee, and that documents have been received.*

Reason for call
- Touching base before the day
- Confidentiality rules in operation (during call)
- Need to outline the process? Especially about opening statements *(depends upon their experience)*
- Role of mediator
- Mediation agreement approved? *(assumes already sent out with the confirmatory email)*

Why are they coming?
- Court scheme?
- Other scheme?
- Is there any resistance?
  - Outline opportunity/no settlement at any cost

Who attends?
- Lawyer? If not, line of contact? If so, then role of supporter
- If no lawyer then partner/companion if party is alone
- Keep team small (and keen)

What has happened so far?
- Settlement discussions/offers?
- Current state of court case/arbitration (if any). Future timetable
- Costs to date/through trial/arbitration?
- Publicly funded?

How might it settle?
- Barriers preventing settlement to date?
- Any limits of authority? Need for out-of-hours contact?
- How might it settle?
- Risk analysis done? What type?

On the day
- Be on time/arrive early
- Any special needs?
- Confirm Mediator contact details
- Strategy for idle time

Anything else?
- Documents/breakdown of quantum
- Settlement agreement pro-forma
- Assistant

Typical Mediation Agreement

There are many forms of mediation agreement (i.e. the parties' agreement to mediate). Some are complex, with several pages of explanatory notes. I favour a simple agreement that is easily understood and that sets down the principles of the mediation.

Date: **Time:**

Venue:

Parties:

(1) Claimant:

(2) Defendant:

Mediator: **Assistant Mediator:**

The Dispute:
(Brief description of the key issues in dispute)

The Agreement:
The undersigned parties hereby agree to participate in mediation in accordance with the following terms:

1. Whilst it is recognised that mediation is a voluntary process and that the mediator will not and cannot compel the parties to settle, nor even to continue negotiating, the parties agree to participate in negotiations in good faith with the aim of achieving a settlement.
2. The parties agree to have present at the mediation such persons as are authorised to agree settlement terms.

3. The parties agree to keep confidential:
   ○ all information, whether oral, written or otherwise, produced for or at the mediation, and
   ○ the terms of any settlement agreement arising from it;
   provided that nothing in this clause prevents the parties (including the mediator) discussing the mediation with the parties' professional advisors and/or insurers and those necessary to implement or enforce the settlement agreement and/or making disclosure to any relevant authority or person, whether under the Proceeds of Crime Act 2002 and/or under any Regulations relating thereto, if obliged to do so by law.
   Note: Evidence that is otherwise admissible or discoverable shall not be rendered inadmissible or non-discoverable simply as a result of its use in the mediation.

4. The parties agree that:
   ○ all offers, promises, conduct and statements made in the course of the mediation proceedings are inadmissible in evidence in any subsequent litigation or arbitration;
   ○ any agreement reached at or following the mediation shall not be binding on the parties unless it is recorded in writing and signed by the parties and/or their authorised representatives;
   ○ they will not call the mediator nor any co-mediator nor any assistant mediator as a witness, nor require the production of records or notes relating to the mediation;
   ○ no recording or transcript will be made at the time of the mediation.

5. The mediation will terminate when:
   ○ a settlement has been reached
   ○ a party withdraws
   ○ the mediator retires for any reason provided by the Chartered Institute of Arbitrators Code of Conduct.

6. Neither the mediator nor any co-mediator nor assistant mediator shall be liable to the parties for any act or omission in connection with the services provided.

7. This agreement will be governed by English law.
   Note: The referral of the dispute to mediation does not affect any rights that may exist under Article 6 of the European Convention on Human Rights. If the dispute is not settled by mediation, the parties' rights to a fair trial are unaffected.

Signed: Date:

Typical Settlement Agreement

Many mediators and lawyers have devised their own template for a settlement agreement. This can be very useful and save time on the day although all agreements are unique and require their own particular clauses. If nothing else the templates provide a useful checklist when drafting a settlement agreement. The one given in the next few pages is used by CEDR (Centre for Effective Dispute Resolution):

Model Settlement Agreement and Tomlin Order

Model settlement agreement

Date _____

Parties

_____('Party A')

[Address][1]_____

_____('Party B')

[Address][2]_____

_____('Party C') etc.]
(jointly 'the Parties')

[Background][3]
The Parties have agreed to settle 'the Dispute' which:

- is being litigated/arbitrated [court/arbitration reference] ('the Action')[4]
- has been the subject of a CEDR Solve mediation today ('the Mediation')

Terms

It is agreed as follows:

1 [A will deliver .... to B at .... by not later than 4 o'clock on 25 December ...][5]

2 [B will pay £......... to A by not later than 4 o'clock on 25 December ... by direct bank transfer to .... bank sort code ... account number ......]

3 ....

4a The Action will be stayed and the parties will consent to an order in the terms of the attached Tomlin Order precedent [see attachment].

OR

4b The Action will be dismissed with no order as to costs.

5 This Agreement is in full and final settlement of any causes of action whatsoever which the Parties [and any subsidiaries .... of the Parties] have against each other.

6 This agreement supersedes all previous agreements between the parties [in respect of matters the subject of the Mediation].[6]

7 If any dispute arises out of this Agreement, the Parties will attempt to settle it by mediation[7] before resorting to any other means of dispute resolution. To institute any such mediation a party must give notice to the mediator of the Mediation. Insofar as possible the terms of the Mediation Agreement will apply to any such further mediation. If no legally binding settlement of this dispute is reached within [28] days from the date of the notice to the Mediator, either party may [institute court proceedings/refer the dispute to arbitration under the rules of ...].

8 The Parties will keep confidential and not use for any collateral or ulterior purpose the terms of this Agreement [except insofar as is necessary to implement and enforce any of its terms].

9 This Agreement shall be governed by, construed and take effect in accordance with [English] law. The courts of [England] shall have exclusive jurisdiction to settle any claim, dispute or matter of difference which may arise out of, or in connection with this agreement.[8]

Signed

_____

for and on behalf of[9] _____

for and on behalf of[10]_____

Note: This Model Agreement, and attached precedent order, is for guidance only. Any agreement based on it will need to adapted to the particular circumstances and legal requirements of the settlement to which it relates. Wherever possible any such agreement should be drafted/approved by each party's lawyer. Although the mediator is likely to be involved in helping the parties to draft acceptable terms, the mediator is not responsible for the drafting of the agreement and does not need to be a party to it.

[See also provisions of mediation agreement which, if it is based on the CEDR Model Mediation Agreement, will deal with mediator liability, confidentiality, etc. and should not need to be repeated in this agreement.]

Attachment to model settlement agreement

Tomlin Order Precedent

[Action heading]

UPON hearing .....

By consent

IT IS ORDERED that all further proceedings in this case be stayed upon the terms set out in the Settlement Agreement between Parties dated ....., an original of which is held by each of the Parties' solicitors [or CEDR Solve/the Mediator] except for the purpose of enforcing the terms of that Agreement as set out below.

AND IT IS FURTHER ORDERED that either Party/any of the Parties may apply to the court to enforce the terms of the said Agreement [or to claim for breach of it] without the need to commence new proceedings.

AND IT IS FURTHER ORDERED that [each Party bear its own costs].

WE CONSENT to an order in these terms

_____[Black & White], Claimant's Solicitors

_____[Red & Green], Defendant's Solicitors

[1] Not strictly necessary

[2] Not strictly necessary

[3] Not strictly necessary but may be useful for setting up definitions

[4] Omit this wording and paragraph 4 if there are no court proceedings

[5] Be as specific as possible, for example, how, by when, etc.

[6] Only necessary if there have been previous agreements

[7] Alternatively, negotiation at Chief Executive level, followed by mediation if negotiations do not result in settlement within a specified time

[8] Usually not necessary where parties are located in same country and subject matter of agreement relates to one country

[9] Not necessary where the party signing is an individual

[10] Not necessary where the party signing is an individual

Reproduced with permission, courtesy of CEDR

Mediation Providers

At the time of publishing there were 40 commercial mediation providers accredited by the CMC (Civil Mediation Council). Accreditation means that the providers use only properly trained mediators and that they have formal complaints processes, minimum insurance cover and mediator monitoring systems. The 40 are:

1. Clerksroom
   Contact name: Stephen Ward
   Equity House, Blackbrook Park Avenue, Taunton, TA1 2PX
   Tel: 0845 083 3000
   Email: ward@clerksroom.com
   Website: www.independentmediators.net

2. Devon & Exeter Law Society
   Contact name: Jeremy Ferguson
   Suite 5, Renslade House, Bonhay Road, Exeter, EX4 3AY
   Tel: 01392 411585
   Email: mediation@devonlawsociety.org.uk
   Website: www.devonlawsociety.org.uk

3. Solent Mediation Limited
   Contact name: Roger Salvetti
   Kingston Place, 62–68 Kingston Crescent, North End, Portsmouth, Hampshire, PO2 8AQ
   Tel: 023 9266 0261
   Email: rsalvetti@biscoes-law.co.uk
   Website: www.solentmediation.com

4. Specialist Mediators LLP
   Contact name: Robin Bryant
   Trinity House, 91 Tarrant Street, Arundel, West Sussex, BN18 9DN
   Tel: 01903 882900
   Email: rbryant@specialistmediators.org
   Website: www.specialistmediators.org

5. Midlands Mediation
   Contact name: Mike Talbot
   Suite 3F, East Mill, Bridgfoot, Belper, Derbyshire, DE56 1XH
   Tel: 01773 829982
   Email: admin@ukmediation.net
   Website: www.ukmediation.net

6. LADR Ltd (Lamb Building Alternative Dispute Resolution Ltd)
   Contact name: Paul Randolph
   Ground Floor, Lamb Building, Temple, London, EC4Y 7AS
   Tel: 020 7797 7788
   Email: ladr@lambbuilding.co.uk
   Website: www.lambbuilding.co.uk

7. Intermediation
   Contact name: John Gunner
   Level 7, Tower 42, 25 Old Broad Street, London, EC2N 1HN
   Tel: 020 7877 0370
   Email: johng@inter-resolve.com
   Website: www.inter-resolve.com

8. Talk Mediation Ltd
   Contact name: Suzanne Lowe
   19-21 King Street, Hereford, HR4 9BX
   Tel: 01432 344666
   Email: enquiries@talkmediation.co.uk
   Website: www.talkmediation.co.uk

9. Global Mediation Ltd
   Contact name: Adam Gersch
   Constable House, Bulwer Road, Barnet, EN5 5JD
   Tel: 020 8441 1355
   Email: info@globalmediation.co.uk
   Website: www.globalmediation.co.uk

10. Littleton Dispute Resolution Services Ltd
    Contact name: David Douglas
    3 Kings Bench Walk North, Temple, London, EC4Y 7HR
    Tel: 020 7797 8600
    Email: david@littletonchambers.co.uk
    Website: www.littletonchambers.co.uk

11. In Place of Strife
    Contact name: Mark Jackson-Stops
    IDRC, 70 Fleet Street, London, EC4Y 1EU
    Tel: 020 7917 9449
    Email: stops@mediate.co.uk
    Website: www.mediate.co.uk

12. ADR Chambers (UK) Limited
    Contact name: Ian Duggan
    City Point, 1 Ropemaker Street, London, EC2Y 9HT
    Tel: 0845 072 0111
    Email: duggan@adrchambers.co.uk
    Webiste: www.adrchambers.co.uk

13. CEDR (Centre for Effective Dispute Resolution)
    Contact name: Graham Massie
    IDRC, 70 Fleet Street, London, EC4Y 1EU
    Tel: 020 7536 6000
    Email: info@cedr.co.uk
    Website: www.cedr.co.uk

14. Mediation-1st
    Contact name: Martin Plowman
    73 The Close, Norwich, Norfolk, NR1 4DR
    Tel: 01603 281128
    Email: martin@mediation-1st.co.uk
    Website: www.mediation-1st.co.uk

15. ADR Group
    Contact name: Michael Lind
    Grove House, Grove End, Redland, Bristol, BS6 6UN
    Tel: 0117 946 7180
    Email: mike.lind@adrgroup.co.uk
    Website: www.adrgroup.co.uk

16. Association of Northern Mediators
    Contact name: Anthony Glaister
    Protection House, 16–17 East Parade, Leeds, LS1 2BR
    Tel: 0113 399 3435
    Email: anthonyglaister@keeblehawson.co.uk
    Website: www.northernmediators.co.uk

17. The Academy of Experts
    Contact name: Nicola Cohen
    3 Gray's Inn Square, London, WC1R 5AH
    Tel: 020 7430 0333
    Email: nac@academy-experts.org
    Website: www.academy-experts.org

18. Mediating Works Ltd
    Contact name: Caroline Buchan
    9 Savill Road, Lindfield, West Sussex, RH16 2NY
    Tel: 0844 477 8022
    Email: c.buchan@mediatingworks.org
    Website: www.mediatingworks.org

19. Oxfordshire Mediation Group
    Contact name: Russell Porter
    3 Paper Buildings, 1 Alfred Street, Oxford, OX1 4EH
    Tel: 01865 793736
    Email: oxford@3paper.co.uk
    Website: www.3paper.co.uk

20. Northern Dispute Resolution
    Contact name: Ronald Bradbeer
    1st Floor, 1 St James Gate, Newcastle upon Tyne, NE1 4AD
    Tel: 0191 233 9785
    Email: advice@ndr-northernmediations.co.uk
    Website: www.ndr-northernmediations.co.uk

21. Maritime Solicitors Association
    Contact name: Rhys Clift
    c/o Hill Taylor Dickinson, Irongate House, 22–30 Dukes Place,
    London, EC3A 7HX
    Tel: 020 7283 9033
    Email: rhys.clift@ltd-london.com

22. Association of Midland Mediators
    Contact name: John J Marshall
    West Midlands: One Victoria Square, Birmingham, B1 1BD
    East Midlands: Cumberland Court, 80 Mount Street, Nottingham,
    NG1 6HH
    Tel: West Midlands: 0121 210 7256

Tel: East Midlands: 0115 936 9368
Email: john.marshall@echarris.com
Website: www.ammediators.co.uk

23. Consensus Mediation Limited
    Contact name: John Winkworth-Smith
    82 King Street, Manchester, M2 4WQ
    Tel: 0870 240 5531
    Email: jws@consensusmediation.co.uk
    Website: www.consensusmediation.co.uk

24. DRS-CIArb (Dispute Resolution Services at the Chartered
    Institute of Arbitrators)
    Contact name: Gregory Hunt
    International Arbitration and Mediation Centre,
    12 Bloomsbury Square,
    London, WC1A 2LP
    Tel: 020 7421 7444
    Email: ghunt@arbitrators.org
    Website: www.arbitrators.org

25. Resolex Limited
    Contact name: Edward Moore
    70 Fleet Street, London, EC4Y 1EU
    Tel: 020 7353 8000
    Email: edward.moore@resolex.com
    Website: www.resolex.com

26. Bristows
    Contact name: Sally Field
    3 Lincoln's Inn Fields, London, WC1A 3AA
    Tel: 020 7400 8000
    Email: sally.field@bristows.com
    Website: www.bristows.com

27. Wandsworth Mediation Services Ltd
    Contact name: Bill Wakely
    61 Wandsworth High Street, London, SW18 2PT
    Tel: 020 8812 4747
    Email: wmediation@btconnect.com
    Website: www.mediationcentre.org

28. Core Mediation
    Contact name: John Sturrock QC
    Rutland House, 19 Rutland Square, Edinburgh
    Tel: 0131 221 2520
    Email: john.sturrock@core-solutions.com
    Website: www.core-solutions.com

29. LawWorks Mediation
    Contact name: Lavinia Shaw-Brown
    10–13 Lovat Lane, London, EC3R 8DN
    Tel: 020 7929 5601
    Email: lsb@lawworks.org.uk
    Website: www.lawworks.org.uk

30. Sports Dispute Resolution Panel
    Contact name: Susan Humble
    Francis House, Francis Street, London, SW1P 1DE
    Tel: 020 7854 8590
    Email: resolve@sportsdispute.co.uk
    Website: www.sportsdispute.co.uk

31. The Association of Cambridge Mediators
    Contact name: John Byrne
    Sheraton House, Castle Park, Cambridge, CB3 0AX
    Tel: 01223 370063
    Email: john.byrne@ccmediation.co.uk
    Website: www.cambridgemediators.co.uk

32. Quadrant Chambers Ltd
    Contact name: Gordon Armstrong
    Quadrant House, 10 Fleet Street, London, EC4Y 1AU
    Tel: 020 7583 4444
    Email: gordon.armstrong@quadrantchambers.com
    Website: www.quandrantchambers.com

33. Middlesex & Thames Valley Mediators (MTVM)
    Contact name: James Torr
    7 Lambscroft Way, Chalfont St Peter,
    Buckinghamshire, SL9 9AY
    Tel: 01753 888 023
    Email: admin@MTVM.org
    Website: www.MTVM.org

34. Immediation Ltd
     Contact name: Anthony Bagshawe
     Widdale Ghyll, Hawes, North Yorkshire, DL8 3LU
     Tel: 0845 257 2734
     Email: info@immediation.co.uk
     Website: www.immediation.co.uk

35. Commercial Mediation West Wales
     Contact name: Kevin O'Brien
     c/o John Collins & Partners LLP, Venture Court,
     Waterside Business Park, Valley Way,
     Enterprise Park, Swansea, SA6 8QP
     Tel: 01792 525445
     Email: post@cmww.co.uk
     Website: www.cmww.co.uk

36. Cornelius Parker Mediation Limited
     Contact name: Vivien Parker
     382-390 Midsummer Boulevard,
     Central Milton Keynes, MK9 2RG
     Tel: 01908 847487
     Email: admin@cpmediation.co.uk
     Website: www.cpmediation.co.uk

37. Wessex Civil Mediators
     Contact name: Paul Cairnes
     30 Christchurch Road, Bournemouth, Dorset, BH1 3PD
     Tel: 01202 292102
     Email: paul.cairnes@3paper.co.uk
     Website: www.3paper.co.uk

38. Mediation Solve
     Contact name: Michael Butterworth
     Suite G11, Central Court, 25 Southampton Buildings,
     London, WC2A 1AL
     Tel: 020 8505 4175
     Email: michael.butterworth@wfg.demon.co.uk

39. BL Resolve
     Contact name: Jonathan Lloyd-Jones
     c/o Blake Lapthorn Tarlo Lyons, Seacourt Tower,
     Westway, Oxford, OX2 0FB

Tel: 01865 248607
Email: jonathan.lloyd-jones@bllaw.co.uk
Website: www.bllaw.co.uk

40. Lyons Davidson Dispute Resolution (LDDR)
    Contact name: Richard Voke
    Bristol House, 40–56 Victoria Street, Bristol, BS1 6BY
    Tel: 0117 904 7744
    Email: info@lyonsdavidson.co.uk
    Website: www.lddr.co.uk

Index

ACAS (Advisory, Conciliation and
    Arbitration Service), 22
added value, 29, 34
adjudication, 17, 18, 20, 23, 26, 28–9,
    31–2, 34, 38, 40, 131, 133
Adj/Med, 20, 26
agenda
    discussions, 86
    facilitation, 129
    timetable, 128
arbitration, 2, 9, 17–18, 20, 23–5, 29, 31,
    32, 34, 36, 38, 40, 74, 83, 94, 96,
    113, 131, 133, 142–3, 155, 160, 161,
    168
arbitrator, 10, 20–22, 23, 38, 78–9, 91, 93,
    99, 105
Arb/Med, 20,26
assistant, 45, 51, 68, 101–2, 155, 156–7
assumptions, 34, 39, 41, 69, 72, 79, 96,
    124
attitude, 12,13, 68, 79, 144
authority, 1
    airport, 6
    decision maker, 56–7, 62, 66, 68, 70,
        93, 101–2, 110, 155, 157
    employer, 19
    granted by parties, 21
    having, 116
    health, 97
    local, 6
    planning, 10
avoiding disputes, 13, 34, 56, 77, 96, 113,
    115, 117, 119, 121, 144

bad faith, 75
BATNA (best alternative to a negotiated
    agreement), 63, 102–3
beauty parade, 48
big picture, 34, 53, 68, 87
blame culture, 11, 27, 57, 107, 114, 118,
    121, 122
bottom line, 83
brainstorming, 79, 88

CCG (Construction Conciliation
    Group), 23, 48, 131
chaired/chairing, 116, 128
Charity Law, 97
chunking up/down, 87
CIF (Construction Industry Federation),
    3, 26
closure, 36
CMC (Civil Mediation Council), 47,
    164
coaching, 34
coal face, 55, 60, 101, 106
collaborating, 13
collaborative negotiation, 1
co-mediation, 47, 50, 65, 77, 109–11
communication, 3, 12, 14, 15, 40, 47, 56,
    67, 87, 96, 101, 110, 115, 121–3
    bad, 12, 15, 115
    effective, 13, 15, 19, 124, 128
    restore, 70
competing, 13
    egos, 50, 110
competitive tendering, 7, 8

conciliation, 18, 19, 22, 23, 48, 131–2
concluding, 135
　negotiating, 41
　stages of mediation, 42, 43, 88, 91, 93,
　　95, 97, 99
confidentiality, 62, 69, 75–6, 79, 99,
　　107–9, 154
　breach of, 22
　provisions, 44–5, 51, 61
confirmatory email, 44, 51–2
conflict, 12–15, 27, 36, 102, 109, 113–20,
　　122, 125
　of interests, 109
confrontation, 108
consensual, 17–19, 147
　process, 19, 22, 27, 147
consensus-building, 103, 131
Consent order, 95
consultants, 3, 5, 6, 8, 9, 11, 56, 101, 106–8
contract, 2–6, 19, 28, 32, 39, 71, 96–7,
　　120–2, 127, 133
　back-to-back, 5
　contract scenario, 4
　disagreement, 3
　interpretation, 5
　no contract scenario, 3
　one-sided, 5
　reasonable standard, 6
　standard contracts, 4–5
　unrealistic criteria, 6
cooling off period, 97–8, 100
cooperating, 102, 113
cost, 1, 2, 7–8, 10, 24–5, 32, 38, 45, 47–9,
　　50–51, 60–61, 82
　consequential, 2
　cost-effective, 9, 25, 28, 110, 133,
　　144
counsel, 26, 37, 52–3, 55–56, 71, 101
court settlement procedure, 20
CPR (Civil Procedure Rules), 19, 26
credible zone, 81–2

culture, 3, 7–11, 15, 27–8, 30, 57, 60, 114,
　　118, 121–22

day in court, 37, 58, 79, 102, 137
dead cert cases, 34
deadlock, 19, 67, 76, 87, 119, 124
deal mediation, 20, 27, 123, 125
deals
　better, 29, 123, 125, 136
　doing a, 6, 8, 30, 31, 35, 36, 39, 40, 47,
　　53, 56, 63, 81, 85, 86, 88, 90, 93, 99,
　　103, 108, 125, 129
　part, 92
　with dignity, 91, 92, 99
decision
　maker, 55–7, 66, 93, 101–3, 106–7,
　　110–11, 135
　point, 113
　tree, 63, 64, 151–3
debrief, 109
default provisions, 96
demonise, 70
dignity, 36, 56, 88, 91–2, 99
dispute avoidance, 1, 27–8, 115, 126
documents, 2, 52, 54, 68, 69, 135, 142–3,
　　154–5
　key, 45, 69
　supporting, 44, 53, 60
DRB (Dispute Resolution Board), 21,
　　28
drivers, 38, 65, 67, 69, 72, 74, 76, 119
dry run, 58–9, 71

emotional intelligence, 39
emotions
　acknowledgement, 39, 102
　parties, 70
ENE (early neutral evaluation), 18, 21,
　　94, 142–3
enquirer, 14
entrenched positions, 35–6

experts, 6, 26, 32–33, 55, 59, 77, 97, 99,
    105–6, 108, 142–3, 150, 167
exploring, 14, 42–3, 67, 73, 74, 76
external factors, 3, 10–11
extreme zone, 82, 84

face (loss of), 34, 36
fairness, 91
    and justice, 30
fight or flight, 114
finality, 8, 26, 33, 40, 91
finance, 3, 7
flipchart, 45, 79
forgiveness, 118, 122
fragmentation, 3, 9;
framework,
    joint and private meetings, 19, 41,
        124, 128
    legal, 103, 105
    typical, 41–2, 61, 70, 120

good faith, 39–40, 62, 69, 75, 86,
    142
government
    departments, 57, 141
    legislation, 11
    negotiations, 57
    structure, 57

headline, 14, 117–18, 122, 129, 130
heads of agreement, 95, 132
history days, 33
Housing Grants and Regeneration Act,
    8, 17

ICE (Institution of Civil Engineers), 22
idle time, 50, 76–77, 80, 103, 110, 155
IDRC (International Dispute Resolution
    Centre), 2
imposed solution, 30, 133
information gathering, 75

initial/opening joint meeting, 33, 68–72,
    106
innovation, 11, 114, 118, 122
insult zone, 82

joint meeting (*see initial/opening joint
    meeting*)
judicial appraisal, 21

Latham Report, 17
law, summary of relevant, 141, 143, 145,
    147
legal argument, 4, 30, 37, 45, 55, 105,
    133, 136
legal costs, 2, 31–2, 46, 82, 83, 96,
    131
legislation, 3, 11
listening, 38, 47, 114, 118, 131, 136
litigation, 9, 17, 18, 24–5, 29, 31–3, 34, 40,
    63, 71, 74, 102, 113, 131
    risk, 63

management time, 26, 32, 40, 46, 91,
    108
margins,
    low, 7
Med/Adj, 23
Med/Arb, 18, 23
mediation agreement, 40, 44, 51, 61–3,
    68, 98, 99
mediator
    big name, 48
    custodian of the deal, 97
    eyes only paper, 44, 53
    fees, 49
    independent, 47
    providers, 47–8, 62, 98, 164–71
    specialist, 48, 50, 65, 110, 127
mistakes, 8, 114, 117–18, 120, 122
multi-party, 50, 65, 67, 93, 94,
    109–11

NFF (neutral fact finding), 21
needs, 10, 13–14, 19, 26, 28–30, 33–5, 38,
    40, 43, 49, 52–4, 57, 61, 65–70, 72–6,
    78, 80–81, 83, 85–6, 88, 91, 94, 101–3,
    105–6, 110, 114, 115, 117, 119–21,
    124, 128, 130, 136
   and drivers, 65, 67, 72, 76
negotiating, 8, 19, 29, 33, 37, 41–3, 57, 65,
    67, 70, 76, 81, 83, 85–90, 98, 103,
    123–5
negotiation
   negotiating a deal, 6, 8
   tenders, 8
neighbour dispute, 132
non-binding, 21, 22, 39, 98
non-financials, 78, 90

offers, 20, 21, 33–4, 36, 40, 44, 48, 52, 61,
    75–6, 81–2, 84–6, 90–91, 99,
102, 104
Olympics, 127
ombudsman, 18, 23
open book, 116, 121
opening joint meeting (*see
   initial/opening joint meeting*)
opening statements, 41, 58, 71–2
openness, 27, 88, 108, 116, 122
others' shoes, 77

pain–pain, 31, 88
partnering, 27, 28, 116, 121–2,
    126–7
   agreements1, 7
   tenders, 8
personality, 12–13, 128
PFI (Private Finance Initiative) and PPP
   (Public Private Partnership), 127,
   130
PIN (Position, Interests, Needs)
   diagram, 74
point of despair, 87
positional negotiation, 86

pre-mediation
   checklist, 61
   meetings, 33, 59, 60, 61
preparing/preparation, 8, 26, 41–5, 47,
    49, 51, 53, 55, 57–9, 61, 63, 65, 82, 90,
    128, 135–6
principled (negotiation), 84–5, 88
private meetings, 19, 33, 41, 51, 64, 70,
    72, 73, 76–7, 129, 136
professional negligence
   claims, 3, 11, 56, 85, 106
project mediation, 27, 119, 120, 123,
    126–7
provider, of mediation, 47–8, 62,
    98
public bodies, 57, 98

reading-in, 20, 44–5, 49, 61
reality-testing, 34, 97, 104
record keeping, 116
reframing, 34
relationships, 3, 4, 6, 14, 19, 24, 27, 30,
    31, 36, 39, 40, 47, 53, 62, 67, 70, 92,
    96, 108, 113–14, 116, 117, 119–20,
    122, 126–7, 136
responsibility, 1, 11, 14, 15, 43, 57, 92, 95,
    98, 103, 114, 127
RICS (Royal Institute of Chartered
   Surveyors), 26, 132
rights, 35
   based negotiation, 135
   legal, 38, 103, 132
risk, 5, 6, 9, 32, 33, 35–6, 46, 58, 63, 64,
    75, 99, 102, 117, 118, 125
   all-risk contracts, 1, 5
   analysis, 63, 66, 82, 83, 135
   assessment, 93
   avoid, 11
   client, 5
   high risk industry, 9
   risk taking, 114
rule book, 115

salami-slicing, 84
serial users, 48
settlement, 5, 18–22, 26, 30–35, 37, 38,
    40, 41, 44–6, 48, 50–55, 60, 62–8, 70,
    72–4, 76, 78–80, 84–7, 89–91, 93, 94,
    97–100, 102, 104–8, 111, 119, 123,
    131, 133, 136
  agreement, 34, 39, 65, 92, 95, 96, 99
  figure, 61
  negotiated, 22, 29, 38
  pot, 31, 88, 90
  rates, 48, 59, 93
  ratification, 57
speed and economy, 32
stages of mediation, 41, 43, 88
story (tell their/your), 15, 29, 30, 33, 34,
    35, 38, 55, 58, 70, 71, 77, 79, 102, 107,
    108, 111, 136, 137
summarising, 119
summary, 44, 45, 52, 53, 119
surveyor, 31, 56
  chartered, 1, 17, 26, 48, 132
  quantity, 1, 10, 11, 35, 106
SWOT (strengths/weaknesses/
    opportunities/threats) analysis, 63

taxation, 96
  income tax, 65, 97, 99
team, 25, 41, 49, 50, 53, 54, 56–61, 66
  legal, 32, 38, 48, 55
  team-building, 13

tender
  competitive tender, 8
  conditions, 8
  margin, 7
tiered resolution, 133
tit-for-tat, 56
Tomlin Order, 95, 158–61
touchy-feely, 38, 39, 76
treadmill, 35, 36, 57, 94
tribunals, 24
truth, version of, 25, 30, 38, 55, 58, 60,
    72, 79, 83, 106–8, 114, 119, 136–7
turn-key, 130

understanding, 14, 15, 21, 30, 38, 43, 73,
    78, 80, 113–4, 117, 124
unrepresented parties, 97

valuing, 70, 113
VAT (value added tax), 97
venue, 44–5, 54, 67, 128
voluntary process, 19, 70

walk-away point, 63, 83
weather-sensitive, 10
when to mediate, 46
win–win, 27, 30, 31, 88
Woolf reforms, 17
working groups, 77, 87, 129

zone of agreement, 81–2, 85